Cosmos in Collision

The Prehistory of our Solar System and of Modern Man

Table of Contents

Caveats...14

Evolution 17

    The Basic Evolution Time Sandwich................................22

    Soft Tissue in Dinosaur Remains 23

    The Case of Flying Birds 24

    Punctuated Equilibria and Problems with Logic................27

Dating Schemes...30

    Decay Rate Schemes...30

    More on Dinosaur Soft Tissue 31

    Petroglyphs..33

    Stratigraphical Dating Schemes 36

    Greenland Ice Core Dating Schemes 39

Part I: The Primordial System...41

    Discovering the Antique Solar System 41

    When Saturn Met the Sun and its Companion, Jupiter 46

    Summary and Takeaways from this Chapter 52

    Mankind's Purple Dawn 54

    Dancing in the Dark 54

    Small and Weak 56

    Life Under a Brown dwarf Star dwarf 57

    The Mystery of Earth Under Saturn..................................65

    Broken Comets and Star Factories....................................68

Saturn's Birth and the Birth of Planet Earth 73

Saturn as the Master of a Free-Floating Planetary Nebula...... 78

Summary and Takeaways from this chapter 81

The Arrival of Man on Earth.. 84

The Cold Reality Facing Primordial Mankind...................... 84

Land of Eternal Shadow... 85

A Fight for Survival 89

Summary and Takeaways from this chapter 90

The Origin of Modern Man: What Were the Requirements?..... 91

Vendramini's Neanderthals.. 96

Elaine Morgan's Aquatic Ape Theory 103

Summary and Takeaways from this Chapter 107

Eyes for bright worlds and for Dark Worlds........................... 108

Summary and Takeaways from this Chapter 112

The Upper Paleolithic World War 113

Predator Versus Warrior 113

Summary and Takeaways from this Chapter 122

Square/Cube Things, and the Ancient Attenuation of Earth's
Gravity 123

Animal and Human Sizes of Past Ages................................ 123

The Weightlifter and the Witch.. 128

How Much Attenuation of Gravity was Needed to Account for
Dinosaur Sizes?.. 130

Sauropod Necks and the Problem of Torque 131

Ancient Flying Creatures 134

Ancient Humans and their Sizes 138

Antediluvian Lifespans 141

Summary and Takeaways from this Chapter 143

Splash Saltations 144

Summary and Takeaways from this Chapter 158

Part II: The Ganymede Hypothesis 159

The Antique Solar System 160

The Strange Case of 'HD 141937 b' 167

Ancient Jupiter as a Glowing Sub-Brown Dwarf Star 169

Brown Dwarfs as Heavenly Water Carriers 173

Other 'Jupiters' that show how our Jupiter Might Once Have Been 175

Basking in the Warmth of a Sub-Brown Dwarf Star............ 177

The Radiation Issue... 181

Summary and Takeaways from this chapter 182

Ganymede: Third Rock from Jupiter 184

Back to the Jovian Moons 184

The Enigma of Ganymede's Moment of Inertia 194

A Ball of Pumice Wrapped Around an Iron Core................ 198

Ganymede's Electrical Conductivity: It's in the P-Holes 200

An Electrifying Scenario: The Birth of Jupiter's Moons 202

Summary and Takeaways from this chapter 205

What Happened to Ganymede?...207

    Frozen Remnants of an Antique Solar System 211

    The Electrical Scarring of Ganymede 213

    Summary and Takeaways from this chapter 226

Club Ganymede: The Antique Solar System's Tropical Paradise
....228

    A Day on Ganymede...229

    Desert Islands: Rocky Lumps and Minerals in the ice........232

    The Dust of Land and Life: Clay and Organic Compounds .236

    Postcards from Ganymede's Lost Atmosphere and Climate 241

    Water, Water Everywhere… and Plenty to Drink.................247

    The Gravity of Life on Ganymede 250

    Summary and Takeaways from this chapter 255

Stellar distances, Friar Occam, and Ganymede 258

Cro-Magnons and Bible Antediluvians....................................260

    Summary and Takeaways from this Chapter 270

Conclusions and Stray Thoughts..271

Selected Bibliography 274

    (Books)...274

    (Selected PDFs, etc.)..276

Appendix A, Telepathy and Pre-Flood Language......................278

    Isolate Languages...278

    Back to the question of Cro-Magnon people. 281

The Evidence for Telepathy in Today's World..........................282

Telepathy in the Ancient World..................................283

Appendix B, Mars 289

Faces...289

Pyramids...293

Cities 294

Phobos...297

Other Artificial Structures...297

Reasons for Denial 299

Forward by Theodore A. Holden

This book is about our ancient solar system, which turns out to have originally been in two parts. One part consisted of our present sun, Jupiter, Mercury, and possibly other bodies that are no longer in evidence. The other consisted of Saturn, which at that time amounted to what we now would call a sub-Brown-dwarf star, along with bodies orbiting Saturn at that time. These included Earth, Mars, and others.

This new vision of cosmology, or at least of our own local cosmology, arises in equal parts from studies of mythology, ancient literature, and iconography, and the findings of plasma physics, which more often than not are at odds with the belief systems of establishment, gravity–only cosmologists. Readers who have kept up with the rise of this neo–catastrophist vision of prehistory over the past several decades will still find much that is new in this book, particularly the chapters involving Jupiter's moons. Others will find pretty much everything in this book new and strange, but we have endeavored to see to it that they will find the logic behind the book's claims to be compelling.

This book also represents an attempt to fill in several of the missing pieces of the neo-catastrophist worldview, most particularly the questions of the origin of modern man. Where did modern man originate? Under what circumstances and when did he get here? By what means did he get here? The findings of this book in these areas will almost certainly be at odds with anything that most readers have seen previously.

This book is not meant to be a theological or religious book of any sort. It uses ancient literature, including religious literature, as historical evidence and Bible scholars may view it as providing bits and pieces of missing information in a few areas. However the authors are not in the religion business and are not in competition with any religions or religious organizations. This is basically a book about very ancient history.

Human history, however, cannot be understood without an understanding of the nature of our solar system both present and in the past and the associations of the ancient pantheon gods and goddesses with planets and particularly with planets such as Jupiter and Saturn, which recently were dwarf stars, are primordial. Thus, the word "god" (with a small 'g') turns up here and there in this book in reference to Jupiter or Saturn. This is not meant to be sacrilegious or any sort of an intentional violation of the first commandment. In fact, a careful reading of Plato's dialogues indicates that educated people in ancient times understood the difference between God and the pantheon/astral gods well enough.

Any work such as this one benefits from a huge amount of work that has gone before and from the heroic efforts of others in the neo-catastrophist community over the past six decades. A huge debt of gratitude is owed to Immanuel Velikovsky, the father of modern catastrophism, to a number of his early associates, particularly Alfred De Grazia, and to the scholars and researchers associated with thunderbolts.info, particularly David Talbott, Dwardu Cardona, Ev Cochran, Wallace Thornhill, Donald Scott and others, as well as to scholars and researchers in what you might call the other branch of the neo-catastrophist family that is mainly concerned with chronological reconstructions, including Charles Ginenthal, Gunnar Heinsohn, Emmet Sweeney, Lynn Rose, and others.

Troy McLachlan first came to my attention in the late winter of 2012 from discussions on the thunderbolts.info forum involving Dwardu Cardona's works. I found his Saturn Death Cult website unusually fascinating in that he had actually read one of Cardona's books (similar to reading a 400 page science journal article) and produced something not only comprehensible to the educated layman, but actually imaginative and enjoyable to read. The fact that he had tried to relate Saturn theory to banking conspiracy theories indicated that he did not suffer from the phobia of defending more than one heresy at a time that afflicts most neo-catastrophists (and prevents their dealing with human origin questions).

I had known for some time that a new addition of the little book dealing with dinosaurs and gravity was needed. In the latter half of 2012 I realized that I needed to write a more general book dealing with prehistory, in which dinosaurs and gravity would be one chapter. Most interesting of all, I had determined the general part of our system where modern humans most likely would have originated, but I didn't have anything more precise than that.

It seemed obvious that such a book would have to include a great deal of material about our ancient solar system, particularly Earth during what Cardona and others refer to as the Purple Dawn era. At that point, I suggested to Troy the possibility of co-authoring a book and, fortunately, he agreed to the idea. In the 30 some years in which I've been following the neo-catastrophist movement, Troy is the only other person with whom the idea of co-authoring a book would have even occurred to me, and in the unlikely event that I could have accomplished such a work on my own, it would have taken five or 10 years.

There are still holes that need to be filled in. For instance, there is a question of what if any relationship may have existed between planets and the "auditory hallucinations" that Julian Jaynes associated with pantheon gods and goddesses in his analysis of the Trojan War. There is a question of where precisely the great apes came from, and there are other such questions. We hope nonetheless that this book will render the collection of such holes needing to be filled in, somewhat smaller.

Foreword by Troy D. McLachlan

"For God so loved the *kosmos* . . ."

John 3:16

Cosmos is a word taken from ancient Greek to describe an order or system. It is often translated and rendered as the English word 'world,' a concept used to describe the general physical reality in which we live. Yet the word 'cosmos' means so much more than the mere ordering of physical reality — it also encompasses the thoughts and processes by which we seek to rationalise or understand our place in this reality we call our world. Life, then, and the attempted understanding of it, *is* cosmos. Our *cosmos* is the ever changing system of paradigms by which we approach, perceive and experience life itself.

But what happens when our cosmos collides with some other cosmos, some other way of thinking or experiencing life that is alien to us?

The answer to that question is that our definitions of reality are forever altered by the violence inherent whenever a prevailing cosmos undergoes collision with a new set of ideas. The subsequent creation of new ideas and new perceptions often fuels new means of communication, new ways of interacting that are both tangible and intangible, spiritual and intellectual. Yet, wherever and whenever the collision of two separate cosmos takes place, there is an inevitable period of grinding and chaotic conflict. Here is the place where accepted and comfortable perceptions that come to define our understanding of life itself are challenged. When such a collision occurs, it not only impacts our sense of the *now*, but also impacts our awareness of the past and the hopes embodied by our views of the future.

Cosmos' in collision can, therefore, be a terrifying, stupefying and even exhilarating experience. It is an Event in the truest sense of the word, a fundamental challenge to the way we see our world, and few of us are ever lucky enough to experience or understand its thrill. Only the brave will choose to stand where two cosmos' collide.

About 10,000 years ago the cosmos governing humanity collided with something that forever changed *the* Cosmos. Human expression changed forever; thinking and art, politics and religion, all changed forever; the environment changed forever; the very physical nature of human existence changed . . . forever!

11

What happened?

It seems a new *Sun* happened . . . and suddenly!

Today we live with the legacy of this once new sun, a sun that has now faded into the realm of forgotten gods and dreamlike memories. But the legacy of this once mighty sun deity is still somehow with us, an ancient inheritance waiting to be discovered deep within our darkest fears and anxieties. This legacy pulls at us, insisting on its place as our first cosmos, *the Cosmos*. It is a legacy that speaks of a former time of great human discovery and agony, a time lost for a thousand years under a mountain of forgotten memories, false ideas and bad science. It is with these same long forgotten memories, human memories, which the current cosmos in which we live today is now beginning to collide. . .

When Ted Holden suggested that we collaborate in writing a book dealing with the pre-history of the Solar System, I was already comfortable with the idea of human origins having its genesis solely on this planet. Flush with the knowledge that mankind's earliest recorded memories talk of a dark primordial time before the arrival of our current Sun, I subscribed wholeheartedly to a cosmology that placed human origins fully within an environment dominated by the dull glow of Earth's original primordial star, a sub-brown dwarf that would later become the planet Saturn. Such a concept was already sufficient to place any proponent on the extreme fringe of accepted scientific discussion, so when Ted suggested that we may need to look for a *bright* world *pre-dating* this era of primordial darkness to fully account for human origins, I found myself facing a crisis of overloaded and seemingly competing paradigms — it seemed to me that Ted was suggesting we leave the relative comfort of the fringe and blast off into uncharted territory clearly marked '***Here be dragons***'.

However, Ted's thinking was original and it highlighted issues important enough to literally force me off-world and to go looking elsewhere in the universe for answers. Armed with Ted's insights, things began to fall into place — almost alarmingly so — and a highly original hypothesis began to take shape. More importantly, a sceptical approach on my behalf failed to overturn the basic premise we were exploring; the off-planet origins of the human species. I had the singular experience that the resulting cosmology, as found in this work, almost began to write itself, as if able to answer its own questions with each new leg of the investigation. Not only that, but it was fun to write in the bargain.

Ted and I have only begun to scratch the surface of what we present in *Cosmos in Collision*. This journey promises to yield much, much more and, as Ted has already pointed out in his own foreword, much is owed to those who have gone before us. In discussions with Ted during the writing of this work, we have both often commented on the sheer enjoyment we have experienced in pursuing this line of enquiry; both of us only hope we have succeeded in relating that same sense of discovery and excitement over the following pages.

Caveats

This book comes with a number of caveats.

This book involves a claim that the planet Saturn was very recently a dwarf star and that Earth and Mars were originally amongst its satellites. A quick Google search on "Saturn Theory" should convince most readers (simply the number of hits) that this idea is not new. This book provides some documentation for this "Saturn Theory" but, other than that, generally just assumes it. An encyclopedic defense of the Saturn theory already exists in the form of Dwardu Cardona's "God Star" (http://www.amazon.com/God-Star-Dwardu-Cardona/dp/1412083087), and the authors of this book are not interested in reinventing wheels.

As the webpage notes, this book involves a startling claim: that the authors have determined to within a statistical certainty the original home of modern humans within our solar system. Also explained is the multi-thousand-year time lapse between the Cro-Magnon saltation and Genesis. The most important thing which we DON'T claim to know is the exact manner in which both of those saltations of modern humans arrived on or were brought to this planet

Neither of the two authors of this book have any prior experience with co-authored books. That can make the flow of the book a tiny bit clumsy in a few places; where one of us wishes to speak of some personal experience or mention the other by name, we'll indicate this by noting "[Troy speaking]" or "[Ted speaking]."

Likewise because of this book being co-authored by two people on opposite sides of the "pond," it contains both British and American spelling variants.

The authors view the theses and conclusions derived in this book to be correct; nonetheless nobody owns the time machine that would be required to absolutely verify any of the conclusions that this book suggests.

In the absence of time machines, this work makes heavy use of the logical principle called "Occam's razor." Named after Friar William of Occam, the principle is generally understood to mean that of competing theories with equal explanatory power, the simplest should be preferred. In particular, given the immense distances between stars in our galaxy, in the presence of a completely plausible origin for modern man within our own solar system, theories involving 'saltations' from other star systems are ruled out.

Virtually all work in the area of natural history over the past century and a half has been within the general paradigm of evolution. It is not possible for all such work to be rubbish or for evolution to totally wreck all good logical thinking over such a space of time. To make sense of a number of these kinds of findings, more often than not, you have to be able to separate the wheat from the chaff. In particular, the works of three scholars who have accepted the general evolutionary paradigm will be examined with an eye towards the logical conclusions that their works indicate if the evolutionary paradigm is removed. These three are Julian Jaynes, a psychologist and philologist who taught at Princeton University and whose "Origin of Consciousness" has remained an academic sensation since the mid-1970s; Elaine Morgan, the most noted advocate of the Aquatic Ape hypothesis; and Daniel Vendramini, a New Zealand scholar who has provided us with the best description and illustration so far of the Neanderthal. Vendramini's thesis of Skhul/Qafzeh hominids morphing into Cro-Magnon people via a fast process of evolution driven by predation from Neanderthals will also be examined.

One of the theses of this book involves static electricity and the question of telepathic communications in prehistoric times. This could possibly suggest certain kinds of dangerous experiments which, for a number of reasons, we view as fundamentally incapable of producing any useful results; we strongly recommend that such experiments not be attempted.

No saltation theory is ever going to save the theory of evolution or do anything more than kick the evolution-vs-design can down the road a block or two; the laws of mathematics and probability work the same way everywhere in the universe as they do here. Evolution remains fundamentally incompatible with modern mathematics, probability theory, and logic.

The term "Cro-Magnon" has been dropped from most scientific literature due to questions as to who or which groups the term should include and we are not comfortable with the idea of including any hominid groups in whatever is supposedly meant by "early modern humans." We generally use the term Cro-Magnon where authors 200 years ago would have written "pre-Adamites," and this is because we're not entirely certain how many distinct saltations of modern humans there have been.

Where other authors and researchers are cited, quoted, referenced or discussed, there is no implication that those authors in any way agree with or endorse the claims and arguments put forward by the two authors of this work.

Evolution

This book is not intended to be an anti-evolution text; there are enough of those available, and they are sufficiently easy to find. Nonetheless this book diverges sharply from standard evolutionary paradigms and a few obligatory pages are in order to describe our reasons for proceeding in such a manner.

Readers who have never believed or have already stopped believing in evolution may want to skip this chapter.

The word "Paradigm" indicates a super theory or general theory (or worldview) that underlies a particular group of studies or scholarly works. As noted above, the theory of evolution amounts to such a general paradigm under which nearly all work in the areas of science formerly called Natural History have proceeded for the last century and a half. This includes nearly all studies of ancient humans and hominids, nearly all studies of dinosaurs, all of the standard dating schemes that we read about and the assumptions upon which they are based, and accepted notions of what kinds of things are possible and what kinds of things are not.

What the educated layman is largely unaware of is the extent to which the theory of evolution has been debunked over the last six or seven decades. Moreover, this unawareness is not accidental; it arises from deliberate policies of government, media, and entrenched academic interests. Most people are also largely unaware that there are two general flavors of evolution, that is, microevolution and macroevolution. Microevolution does not involve any controversies; it amounts to variation within a particular kind, or species, that allows creatures of that species to adapt to a changing environment. Examples include finches with one kind of beak morphing into finches with a different kind of beak, Brown moths morphing into white moths and that kind of thing. Microevolution can also account for the difference between a leopard and a jaguar.

But microevolution is not what the theory of evolution is about. The theory of evolution is about macroevolution. That is, it is about the idea that microevolutionary changes could agglomerate into a change from one kind of creature to another, with new organs, new requirements for system integration between new and old organs, and a new basic plan for existence. This is the idea that has been debunked, and the debunkings are sufficient in number and sufficiently independent, one from the others, that they cannot be ignored. The theory survives today largely due to inertia, and to the fact that it serves as a latter-day religion to its adherents.

The first such debunking involved the common fruit fly. Fruit flies breed new generations every few days. If you run experiments on fruit flies for several **DECADES**, as has been done starting in the early 1900s, then the experiments will involve more generations of fruit flies than there have ever been of humans or of anything at all resembling humans on this planet. What the fruit fly experiments represented was a controlled laboratory test of the theory of evolution, which attempted to *evolve* new animal species via mutation as evolution requires.

These flies were bombarded with everything known to cause mutations; chemicals, every kind of radiation, extreme temperatures, blast, shock, noise, etc. Attempts were also made to recombine mutants, and all they ever got was precisely what the breeders had told Darwin would be all he'd ever get, i.e. sterile individuals, and individuals that returned, boomerang-like, to the norm for a fruit-fly after two or three generations. Basically, all they ever got were fruit flies. No gnats, no wasps, no ants, no roaches, hornets, spiders, or anything else — just fruit flies[1].

Again, a decades-long experiment like that would be equivalent to hundreds of thousands of years' worth of accumulated genetic change amongst humans or higher animals. The results were so unambiguous that a number of the scientists involved publicly

---

[1] For instance, Kyle Butt, M.A. "Mutant Fruit Flies Bug Evolution", http://www.apologeticspress.org/APContent.aspx?category=9&article=250 1

renounced evolution, the most notable case being that of Richard Goldschmidt. As a consequence, he claimed he was afterwards subjected to treatment similar to the two-minute hate sessions in Orwell's famous novel *1984*[2]. Such is the state of American academia.

Evolution is known to be incompatible with modern mathematics and probability theory. None other than so distinguished a mathematician as Sir Fred Hoyle has noted that:

> "The likelihood of the formation of life from inanimate matter is one to a number with 40,000 noughts after it... It is big enough to bury Darwin and the whole theory of Evolution. There was no primeval soup, neither on this planet nor on any other, and if the beginnings of life were not random, they must therefore have been the product of purposeful intelligence."[3]

But the problem goes beyond numbers, and involves ideas from the realms of the theories of information and computation. The DNA/RNA system, which forms the basis for all life as we know it, amounts to an information system, and information systems do not just happen out of the blue. Notable scientists have commented on this, for instance Charles B. Thaxton, a Ph.D. in chemistry, a postdoctoral Fellow at Harvard and staff member of the Julian Center:

> "...an intelligible communication via radio signal from some distant galaxy would be widely hailed as evidence of an intelligent source. Why then doesn't the message sequence on the DNA molecule also constitute prima facie evidence for an intelligent source? After all, DNA information is not just analogous to a message sequence such as Morse code, it is such a message sequence."[4]

---

[2] Phillip E, Johnson "Darwin on Trial", p. 58
[3] Sir Fred Hoyle, *Nature*, Nov 12, 1981, p. 148
[4] Charles B. Thaxton, "The Mystery of Life's Origin: Reassessing Current Theories", Philosophical Library, 1984, pp. 211-212.

Also adding his weight to the argument was I.L. Cohen, researcher and mathematician, a member of the NY Academy of Sciences, and Officer of the Archaeological Inst. of America:

> "At that moment, when the DNA/RNA system became understood, the debate between Evolutionists and Creationists should have come to a screeching halt."[5]

The failure of the fruit fly experiments was precisely due to the fact that our entire living world is driven by coded information and that the only information that there ever was in the picture was that which fundamentally detailed a fruit fly. When the DNA/RNA information scheme was discovered, even if the fruit fly business had never happened, evolution should have been discarded on the spot. But *GIVEN* the failure of the fruit fly experiments, somebody *had* to have thought to himself:

> *"Hey,* **THAT'S THE REASON THE FRUIT FLY EXPERIMENTS FAILED***!!!!!"*

Again, the DNA/RNA system is an information code just like C#, Java, or C++, and information codes do not just sort of happen or appear amongst inanimate matter for no particular reason.

It gets even worse than that. It turns out that living cells have all of the main components of modern digital computers and computing systems. The main starting point for reading this part of the story will be Donald E Johnson's "*Programming of Life*", and the website and YouTube area associated with Johnson's works:

---

[5] I.L. Cohen, "Darwin Was Wrong - A Study in Probabilities," New Research Publications, 1984, p. 4

http://programmingoflife.com/

http://www.youtube.com/user/programmingoflife

http://www.youtube.com/watch?v=hRooe6ehrPs

Readers won't find Johnson's book easy to read. Nonetheless, you can picture a computer and the way in which it operates by picturing a human performing some of the same kinds of tasks which computers perform. Thus a human pretending to be a computer program would need a chalkboard (like RAM memory) upon which to perform calculations and modify data, input data (another human feeding him numbers or character strings), data storage (a pencil and some paper) upon which to store computational results, and branching logic (his mind) to determine what to do with a particular piece of input data, according to the data itself. That last item, information which directs processes and which Johnson and other computer scientists refer to as "functional information," is the most critical item in the picture. Living cells turn out to have all of those components, including functional information. As Johnson notes, this cellular information system directs the catalyzed production of biological components in milliseconds, which would require more than the theoretical age of the universe to occur without the enzyme catalysts.

The Basic Evolution Time Sandwich

There is also a question of time, and what we refer to as the basic evolution time sandwich, which involves realistic assessments of the amounts of time that evolutionary scenarios need, and that they actually have.

The slice of bread on the bottom of this time sandwich is called the Haldane Dilemma, which amounts to an understanding of the time spans that would be needed to spread *any* genetic change through *any* group of creatures. A very simple version of the concept is all most people should need and can be summed up by the following:

> Imagine a population of 100,000 apes or "proto-humans" ten million years ago. They are all genetically alike other than for two with a "beneficial mutation." Imagine also that this population has the human or proto-human generation cycle time of roughly 20 years.

> Imagine that the beneficial mutation in question is so good, that all 99,998 other apes die out immediately (from jealousy), and that the pair with the beneficial mutation thereafter has 100,000 kids and thus replenishes the herd.

> Imagine that this process continues for ten million years, which is more than anybody claims is involved in human evolution. The maximum number of such "beneficial mutations" that could thus be substituted into the herd would be ten million divided by twenty, or 500,000 point mutations. Walter Remine notes[6] that would be about 1/100 of one percent of the human genome, and a miniscule fraction of the 2 to 3 percent that separates us from

---

[6] Walter Remine, "Biotic Message", page 209

chimpanzees, or the half of that which separates us from Neanderthals[7].

That basically says that, even given a rate of evolutionary development that is fabulously beyond anything that is possible in the real world, starting from apes, in ten million years the best you could possibly hope for would be an ape with a slightly shorter tail. Of course, given more realistic assumptions about the rates at which genetic mutations could be spread through groups of animals, the time required for any kind of an ape to man evolution scenario would be vastly greater than 10 million years.

Soft Tissue in Dinosaur Remains

The slice of bread on top of the time sandwich involves a reasonable assessment of the amount of time that evolutionists actually have for their scenarios. A fairly good idea of the general dimensions of this time limit has been coming in lately in the form of blood, blood vessels, collagen, and raw meat turning up in dinosaur remains, with paleontologists describing dinosaur remains as having a "grave-yard" smell about them. The first such find involved a tyrannosaur leg bone that had to be cut in half in order to be transported via helicopter from the site at which it was found. Other recent finds include a hadrosaur that is described as "mummified." An article in *Scientific American* notes:

> "Unlike most fossils of dinosaur skin, which just result from scales indenting the sediment around them, the finding here contains actual rust-brown mineralized skin, complete with microscopic cell-like structures between five and 30 microns wide. "I'm amazed at the evidence of actual skin structure preserved in the fossil," remarks vertebrate

[7]Fossil DNA proves Neanderthals were not ancestors of humans, http://expressindia.indianexpress.com/fe/daily/19970712/19355423.html

paleontologist Mike Benton of the University of Bristol in England, who did not participate in this study."[8]

Those kinds of organic materials simply don't remain after millions of years[9].

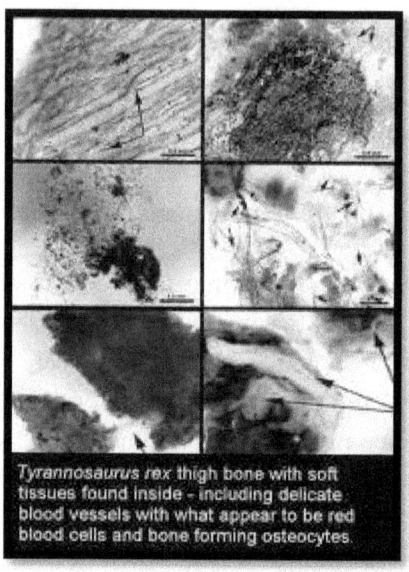

Pictures[10] of organic remains from dinosaurs that we are told have survived millions of years.

The Case of Flying Birds

A common sense grasp of the enormity of the problem can be had by considering what would have to happen for flying birds to evolve.

---

[8] http://www.scientificamerican.com/article.cfm?id=scientists-flesh-out-foss
[9] http://www.sciencemag.org/content/319/5859/33.3.full and http://www.sciencemag.org/content/319/5859/33.3/F1.large.jpg
[10] http://www.detectingdesign.com/fossilizeddna.html

Suppose then for a moment that you are not a flying bird, but that you wish to become one or for your descendants to be flying birds. You'll need a baker's dozen of highly specialized systems, including wings, flight feathers, the system that birds use to rotate flight feathers on up-strokes and down-strokes (the idea being for the wings to open on upstrokes and close on down strokes like a Venetian blind), a specialized light bone structure, specialized flow-through design lungs and an equally high-efficiency heart, a specialized tail, a specialized general balance parameters, and so on.

For starters, every one of these things would be anti-functional until the day on which all such elements came together, so that the chances of evolving any of these things by any process resembling evolution (mutations plus selection) would amount to an infinitesimal, i.e. one divided by some gigantic number.

In probability theory, to compute the probability of two things happening at once, you multiply the probabilities together. That says that the likelihood of all these things ever happening, best case, is ten or twelve such infinitesimals multiplied together i.e. a tenth or twelfth-order infinitesimal. The whole theoretical history of the universe isn't long enough for that to happen even once.

All of that was the best case scenario. In real life, natural selection could not plausibly select for a 'hoped-for' functionality, which is what would be required in order to evolve flight feathers on something that could not fly a priori. In real life, all you'd ever get would be some sort of a random walk around some starting point, rather than the unidirectional march towards a future requirement which evolution requires.

And the real killer is the following consideration: Assume that you were to somehow miraculously evolve the first feature you'd need to become a flying bird. Then by the time another 10,000 generations rolled around and you evolved the second such feature, the first, having been dysfunctional/anti-functional all the while, would have **DE-EVOLVED** and either disappeared altogether or become vestigial.

Consider how close anything comes to such an evolutionary lizard-to-bird process in real life. A coelurosaur trying to evolve its way to bird-status would need a dozen extra systems that it does not have.

Chickens, on the other hand, have all of those things and you might wonder what keeps chickens from ever totally regaining flight. The basic answer is that the chicken as we know it started out as a little two-pound jungle fowl[11] and was bred into a 6-lb. meat animal, but still has the 2-lb bird's wings. Geese are as heavy as chickens and fly easily enough because they have the wings necessary for a 7-lb bird.

Nonetheless, for all of that, chickens lack very little in being what you might call a flying bird. Consider also that man raises chickens in great abundance, and that they were never kept in cages until the 1960s. Consider the numbers of such chickens that must have escaped in all of recorded history; look in the sky overhead: Where are all of their wild-living descendants?? Why are there no wild chickens in the skies above us???

In other words, if there's any chance whatsoever of a non-flying creature evolving into a flying bird, then surely the escaped or feral chicken, close as it is, could **RE-EVOLVE** back into a flying bird. After all, they're only missing the tiniest fraction of whatever is involved. They've got wings, tails, and flight feathers, and the whole nine yards. In their domestic state, they can fly albeit badly; they are entirely similar to what you might expect of an evolutionist's proto-bird, in the final stage of evolving into a flight-worthy condition.

According to evolutionist dogma, at least a few of these should very quickly finish evolving back into something like a normal flying bird, once having escaped, and then the progeny of those few should very quickly fill the skies.

---

[11] http://simple.wikipedia.org/wiki/Red_Junglefowl

But the sky holds no wild chickens. In real life, against real settings, real predators, real conditions, the imperfect flight features do not suffice to save them.

These observations lead to one conclusion: *In real life, if you ever lose the tiniest part of some complex trait or capability, you will never get it back. In the real world, if you lack the tiniest part of some complex trait or capability then, other than possibly via some genetic engineering process, you will never get it.*

The basic question is: how is some dinosaur, having none of the requirements for flight a priori and starting from a miniscule numeric base, supposed to cover the vast evolutionary distance required to become a flying bird, when the domestic chicken, with a numeric base in the billions and lacking only the tiniest bit of what is needed to regain full flight capabilities, cannot even cover that tiny evolutionary distance.

Punctuated Equilibria and Problems with Logic

Real science theories do not require reinvention every couple of decades. Evolutionists, on the other hand, never have to wait terribly long for the next version. Because there is zero evidence in the fossil record to support Charles Darwin's original gradualist concept of macroevolution and because the original conceptions of evolution have been flatly refuted by developments in population genetics since the 1950's, the latest incarnation of evolution theory is Steve Gould and Niles Eldredge's "Punctuated Equilibria" concept, commonly called "*punc-eek*" (PE). This version of the theory of evolution claims that these wholesale violations of probabilistic laws all occurred so suddenly as to never leave evidence in the fossil record. Further, that they all occurred amongst tiny groups of animals living in "peripheral" areas. That is to say, that punctuated equilibria amounts to an attempt to get by both the Haldane dilemma and the problem of the general lack of intermediate fossils between the supposed stages evolving species go through.

Punc-eek amounts to a claim that all meaningful evolutionary change takes place in peripheral areas, amongst tiny groups of animals that develop some genetic advantage, and then move out and overwhelm, outcompete, and replace the larger herds. The claim is that this eliminates the need to spread genetic change through any sizeable herd of animals and, at the same time, explains why we never find intermediate fossils (since there are never enough of these **CHANGELINGS** to leave fossil evidence).

Obvious problems with punctuated equilibria include, minimally:

1. It is a pure pseudoscience seeking to explain and actually be proved by a lack of evidence rather than by evidence (all the missing intermediate fossils). In other words, the people promoting this idea are claiming that the very lack of intermediate fossils supports the theory.
2. PE amounts to a claim that inbreeding is the most significant source of genetic advancement in the world
3. PE requires these tiny peripheral groups to conquer vastly larger groups of animals millions if not billions of times. This is like requiring Custer to win at the Little Big Horn every day, for millions of years.
4. PE requires an eternal victory of animals specifically adapted to localized and parochial conditions over animals that are globally adapted. In real life, the globally adapted animals almost invariably win .
5. For any number of reasons, you need a minimal population of any animal to be viable. This is before the tiny group even gets started in overwhelming the vast herds. A number of American species such as the heath hen became non-viable when their numbers were reduced to a few thousand; at that point, any stroke of bad luck at all, a hard winter, a skewed sex ratio in one generation, a disease of some sort, and it's all over. The heath hen was fine as long as it was spread out over the East coast of the U.S. The point at which it got penned into one of these "peripheral" areas, which Gould and Eldredge see as the salvation for evolutionism, it was all over[12].

The sort of things noted in points 3 and 5 are generally referred to as

---

[12] http://en.wikipedia.org/wiki/Heath_Hen

the "gambler's problem." In this case, the problem facing the tiny group of "peripheral" animals is similar to that facing a gambler trying to beat the house in blackjack or roulette. The house could lose many hands of cards or rolls of the dice without flinching, and the globally-adapted species spread out over a continent could withstand just about anything short of a continental-scale catastrophe without going extinct. On the other hand, two or three bad rolls of the dice will bankrupt the gambler, and any combination of two or three strokes of bad luck will wipe out the "peripheral" species. Gould's basic method of handling this problem is to ignore it.

And there's one other thing that should be obvious to anybody attempting to read through Gould and Eldredge's writings, i.e. they don't even bother to try to provide a mechanism or technical explanation of any sort for these claims. They are claiming that at certain times, amongst tiny groups of animals living in peripheral areas, a "speciation event(TM)" happens, and then the rest of it takes place. In other words, they are saying:

> **ASSUMING that Abracadabra-Shazaam happens, then the rest of the business proceeds as we have described in our scholarly discourse above!**

Again, Gould and Eldredge require that the Abracadabra-Shazaam happen not just once, but countless billions of times, i.e. at least once for every kind of complex creature that has ever walked the Earth. They do not specify whether this amounts to the same Abracadabra-Shazaam each time, or a different kind of Abracadabra-Shazaam for each creature.

Walter Remine has noted[13] that neither Darwinian gradualism nor punctuated equilibria is logically coherent and that evolutionary theorists are now serving up what he describes as a smorgasbord containing bits and pieces of both. Remine's *"The Biotic Message"* is a thoroughgoing analysis of the Haldane dilemma and of related

---

[13] http://www.nmsr.org/round1b.htm

topics.[14] More recently, Remine has produced several papers that provide a simplified description of the dilemma, which do not involve the concept of genetic death.

Dating Schemes

Decay Rate Schemes

Another preliminary idea, which is necessary for this book, involves the logical and scientific schemes that produce and demark the ages and spaces of time that we are accustomed to seeing in any sort of a discussion of dinosaurs, paleontology, or any other facet of pre-historical reality. We noted in the previous chapter that researchers are beginning to find soft tissue in dinosaur remains. This raises an obvious question about the actual antiquity of dinosaurs, as well as a question of what else could be wrong with standard dating techniques.

The schemes and techniques, which scientists use to produce the kinds of dates that we see in journal articles and in the media, depend upon assumptions of uniformity. That is to say, they depend upon the notion that the processes that we observe in the world today are the only ones that there ever could have been. The first minute that you ever allow for any kind of a planetary scale catastrophe to have taken place in ancient times, the logical basis for all such techniques becomes highly suspect.

Radiocarbon dating, for instance, depends on the idea that the ratios of carbon isotopes in our atmosphere have always been what they are now. There is a question, however, of whether even a relatively minor event involving a comet, such as that associated with the Roman Emperor Justinian[15], could permanently change such an

---

[14] http://saintpaulscience.com/index.html

[15] Science Daily "Astronomers Unravel A Mystery Of The Dark Ages," http://www.sciencedaily.com/releases/2004/02/040204000254.htm

isotope ratio. Worse would have been the planetary disaster which Velikovsky and Ignatius Donnelly described[16] and which Greco Roman literature associates with the Phaeton legend, and/or the flood at the time of Noah. Either of those two events could easily have changed such isotope ratios.

Atomic decay dating schemes (uranium/lead dating for instance) assume that the heavy metals that we find near the earth's surface are indigenous to the planet. The problem is that the standard model, which has planets forming from swirling masses of solar material under the influence of gravity, would lead one to expect that heavy metals would all end up near the planet's core. If you assume that heavy metals that we find near the surface got there either via impact events or were generated (as electric universe theorists would have it) by plasma physics phenomena such as arc discharges, then trying to deduce an age for the planet from such metals is questionable.

Not that there is any reason to believe in any sort of 6,000 to 10,000 year age for our planet. We actually have one planet, Venus, in our system that is in fact ballpark for some sort of an age estimate of less than 10,000 years, and Venus *LOOKS* that age[17]: 900°F surface temperature, massive thermal imbalance, massive upwards infrared flux, 90 bar CO_2 atmosphere (a three mile-per-hour wind would knock you over), statistically random cratering, total lack of regolith etc. etc. Since Mars and Earth do not look like that in any way, shape, or manner, you have to assume they are older than that, but not necessarily hundreds of millions or billions of years old. Robert Bass once[18] redid Lord Kelvin's heat equations for the Earth *WITH* a maximal figure for radioactive elements and came up with an upper-bound age of around 200 million years, but that should likely be viewed as an extreme upper bound.

More on Dinosaur Soft Tissue

---

[16] http://www.amazon.com/Worlds-Collision-Immanuel-Velikovsky/dp/1906833117

[17] http://www.bearfabrique.org/Catastrophism/venus/venus2007.pdf

[18] Personal correspondence

Again, we noted in the previous chapter that soft tissue is beginning to be found in dinosaur remains that this is not in keeping with an antiquity which is supposedly in the tens of millions of years. The first such case involved a tyrannosaur leg bone unearthed in Montana and analyzed by scientists at North Carolina State University. The BBC noted of the find (12 April 2007):

> "Researchers compared organic molecules preserved in the *T. rex* fossils with those of living animals, and found they were similar to chicken protein. The discovery of protein in dinosaur bones is a surprise - organic material was not thought to survive this long."

The assumed impossibility of organic materials surviving into the millions of years originally led skeptics to deny the claims:

> "Skeptics argued that the alleged organic tissues were instead biofilm—slime formed by microbes that invaded the fossilized bone."[19]

Nonetheless, an increasing number of such finds and careful techniques to avoid contamination have forced the skeptics to redirect their efforts towards explaining how such materials actually could survive tens of millions of years…

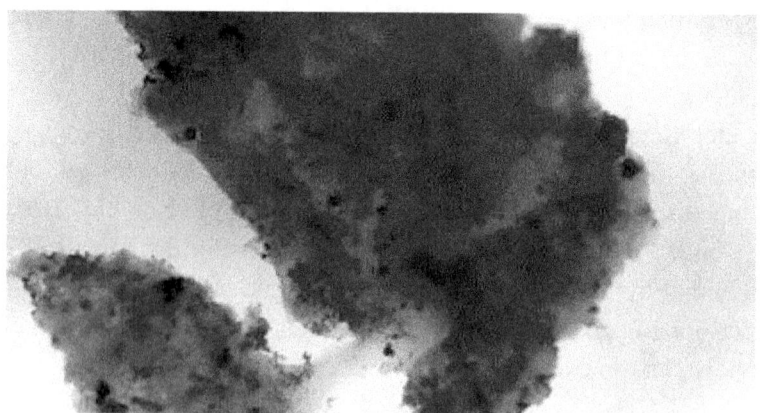

Image[20] of the Tyrannosaur leg bone unearthed in Montana with intact organic soft tissue. For a creature that is supposed to have

[19] http://blogs.scientificamerican.com/observations/2012/10/18/molecular-analysis-supports-controversial-claim-for-dinosaur-cells/
[20] http://www.smithsonianmag.com/science-nature/dinosaur.html

died out millions of years ago, finding intact soft tissue in its remains is a virtual impossibility.

Even more astonishing is the case of a hadrosaur uncovered in 1999 on a North Dakota ranch, which is still being analyzed. Photos obtained by the BBC show clear scales and cross sections of microscopic tendon structures (see image below). The BBC reported[21] that:

> "Tests have shown that the fossil still holds cell-like structures - but their constituent proteins have decayed. The team says the cellular structure of the dinosaur's skin was similar to that of dinosaurs' modern-day descendants."

Petroglyphs

Aside from soft tissue in dinosaur remains, there are recognizable dinosaur images on canyon walls and around lakes and rivers at various North American sites. These are referred to as "petroglyphs" or rock art.

---

[21] news.bbc.co.uk/2/hi/8124098.stm

One very clear depiction of a sauropod dinosaur occurs at Natural Bridges (Kachina Bridge) Utah:

Raw image:

Image with outline amateurishly traced:

Another sauropod image turned up in the 1920s during the Doheney Scientific Expedition to the Hava Supai Canyon in Arizona (see image below).

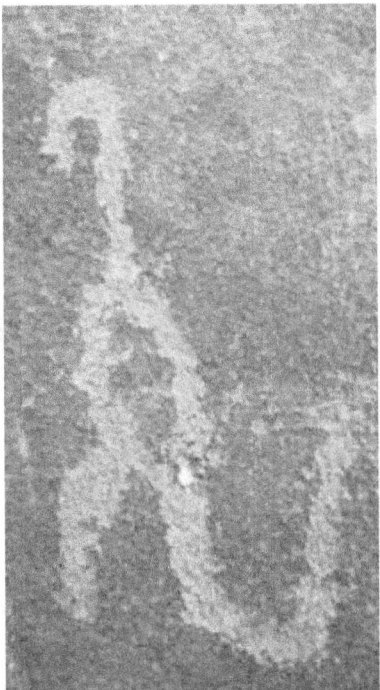

The report from the Doheny Expedition is kept at the Peabody Museum of American Ethnology at Harvard University. A reprint of the report is available on Amazon[22].

Another curious petroglyph can be seen in the following image:

---

[22] http://www.amazon.com/BioFortean-Reprint-Scientific-Expedition-Northern/dp/1616460687

American Indians habitually touched such glyphs up every few decades and the horns were added at a much later date by a painter who assumed the creature needed them. Amerind oral traditions describe this creature, called Mishipishu (the "water panther" in Ojibwa language), as having red fur, a saw-blade back, and a "great spiked tail" that he used as a weapon. Vine Deloria noted[23] that this was basically a description of a stegosaur.

Stratigraphical Dating Schemes

But what about the time frames and scales involving hominids and modern man? We've seen that the time frames generally given for dinosaurs are questionable, but what about the 50,000, 200,000 and 500,000 year time spans you read about for the supposed history of human ancestors (or at least of relatively recent human ancestors)? 50,000 Years and greater is outside the effective range of radiocarbon dating, and many if not most of these numbers arise from studies of stratigraphy.

---

[23] Vine Deloria Jr. "Red Earth, White Lies", pp. 242-243

As an answer to this issue we can turn to the thoughts of one Gunnar Heinsohn of the University of Bremen. Heinsohn is a major player in the efforts to reconstruct the chronologies of the ancient Mediterranean basin and a frequent speaker at NATO gatherings, since his population youth bulge theories predict political unrest in the world with near 100% accuracy. His work "*Wie Alt ist das Menschengeschlect*" describes the problem with the dating schemes typically associated with studies of the Neanderthal. He notes[24]:

> "Mueller-Karpe, the first name in continental paleoanthropology, wrote thirty years ago on the two strata of homo erectus at Swanscombe/England: "A difference between the tools in the upper and in the lower stratum is not recognizable. (From a geological point of view it is uncertain if between the two strata there passed decades, centuries or millennia.)" (Handbuch der Vorgeschichte, Vol I, Munich 1966, p. 293).
>
> The outstanding scholar never returned to this hint that in reality there may have passed ten years where the textbooks enlist one thousand years. Yet, I tried to follow this thread. I went to the stratigraphies of the Old Stone Age, which usually look as follows
>
> modern man (homo sapiens sapiens)
>
> Neanderthal man (homo sapiens neanderthalensis)
>
> Homo erectus (invents fire and is considered the first intelligent man).
>
> In my book "Wie alt ist das Menschengeschlecht?" [How Ancient is Man?], 1996, 2nd edition, I focused for Neanderthal man on his best preserved stratigraphy: Combe Grenal in France. Within 4 m of debris it exhibited 55 strata dated conventionally between -90,000 and -30,000. Roughly one millennium was thus assigned to some 7 cm of debris per stratum. Close scrutiny had revealed that most strata were only used in the summer. Thus, ca. one

---

[24] Personal correspondence.

thousand summers were assigned to each stratum. If, however, the site lay idle in winter and spring one would have expected substratification. Ideally, one would look for one thousand substrata for the one thousand summers. Yet, not even two substrata were discovered in any of the strata. They themselves were the substrata in the 4 m stratigraphy. They, thus, were not good for 60,000 but only for 55 years.

I tested this assumption with the tool count. According to the Binfords' research--done on North American Indians--each tribal adult has at least five tool kits with some eight tools in each of them. At every time 800 tools existed in a band of 20 adults. Assuming that each tool lasted an entire generation (15 female years), Combe Grenals 4,000 generations in 60,000 years should have produced some 3.2 million tools. By going closer to the actual life time of flint tools tens of millions of tools would have to be expected for Combe Grenal. Ony 19,000 (nineteen thousand) remains of tools, however, were found by the excavators.

There seems to be no way out but to cut down the age of Neanderthal man at Combe Grenal from some 60,000 to some 60 years. I applied the stratigraphical approach to the best caves in Europe for the entire time from Erectus to the Iron Age and reached at the following tentative chronology for intelligent man:

-600 onwards Iron Age
-900 onwards Bronze Age
-1400 beginning of modern man (homo sapiens sapiens)
-1500 beginning of Neanderthal man
between -2000 and -1600 beginning of Erectus.

Since Erectus only left the two poor strata like at Swanscombe or El-Castillo/Spain, he should actually not have lasted longer than Neanderthal-maybe one average life expectancy. I will now not go into the mechanism of mutation. All I want to remind you of is the undisputed sequence of interstratification and monostratification in the master stratigraphies. This allows for one solution only: Parents of the former developmental stage of man lived together with their own offspring in the same cave stratum until they died out. They were not massacred as textbooks have it:

monostrat.: only modern man's tools
interstrat.: Neanderthal man's and modern man's tools side by side
monostrat.: only Neanderthal man's tools
interstrat.: Neanderthal man's and Erectus' tools side by side
monotstrat.: only Erectus tools (deepest stratum for intelligent man)

The year figures certainly sound bewildering. Yet, so far nobody came up with any stratigraphy justifiably demanding more time than I tentatively assigned to the age of intelligent man. I always remind my critiques that one millennium is an enormous time span--more than from William the Conqueror to today's Anglo-World. To add a millennium to human history should always go together with sufficient material remains to show for it. I will not even mention the easiness with which scholars add a million years to the history of man until they made Lucy 4 million years old. The time-span-madness is the last residue of Darwinism."

Heinsohn is not putting an exact age on the Neanderthal die-out; what he *is* stating is that there is no legitimate interpretation of existing stratigraphical evidence that would indicate that they occupied the main caves in which their remains are found for more than a few decades. He assumes that hominids progressed to humans via some unknown process that was much faster than evolution.

Greenland Ice Core Dating Schemes

No discussion of dating schemes would be complete without some mention of *Glacier Girl*, the Amalekite P38.

Glacier Girl is a P38 fighter plane which went down over Greenland around 1941. It has recently been recovered and restored to flight-worthy condition. Working from standard theories regarding the rates at which layers of ice and snow accumulate on the Greenland ice shelf, search teams anticipated finding the aircraft two or 3 feet below the surface of the ice and snow. The 260 foot depth at which it was actually found would correspond to it having gone down at roughly the time of the Israelite Exodus from Egypt. In theory, that should make Glacier Girl an element of the Amalekite Air Force, since the only military power on Earth at that time with the economic and financial wherewithal to produce anything like a P38 would have been the Amalekite/Hyksos empire that dominated Egypt after the Middle Kingdom.

The basic reality is that the educated layman needs to learn to take the dates and dating schemes proposed by modern science with several grains of salt. That, and to keep in mind Heinsohn's dictum that a thousand years is an absolutely gigantic space of time.

Part I: The Primordial System

Discovering the Antique Solar System

For many, the Bible and other significant works of ancient literature can be difficult to read because they appear to be describing a number of issues pertaining to the realities of ancient life that were so substantially different from anything experienced today, that they do not seem credible to the modern reader. Modern readers, particularly scientists and scholars, assume that these stories employed fanciful techniques and involved allegorical meanings. They generally do not entertain the possibility that the ancient authors were relating to their readers what amounted to everyday reality in their world, and that they were, in actuality, describing such realities in the idioms of their language and not ours.

There are several broad categories for such things; these include:

- Descriptions of planetary scale catastrophes, including the flood at the time of Noah, the incident associated with the tale of the tower of Babel, the story of Phaeton, and several other tales of a similar nature.
- Descriptions of religious practices intended to communicate directly with the spirit realm. Such practices included prophecy, oracles, "familiar spirits" (the tale of Saul, Samuel, and the "witch of Endor," etc.), idolatry and the rituals associated with the worship of idols, and electrostatic devices such as the Ark of the Covenant. All such practices involved humans undergoing trance-like states similar to hypnosis, all appear to have involved electrostatic phenomena and all stopped working prior to the time of Alexander.
- Indications that the Solar System itself was substantially different in ancient times. The earliest religions were astral in nature and the name associations between our planets and the astral pantheons of gods and goddesses

41

have primordial roots. Primitive people in our present world attempting to devise an astral religion from scratch would invariably end up worshipping the sun and the moon. Nonetheless, the two chieftain gods of every one of those ancient pantheons were Jupiter and Saturn, and not the sun or moon. Plato consistently refers to antediluvians as "nursling's of Kronos (Saturn)"[25] and ancient authors, including Ovid and Hesiod, described the age prior to the flood as having been a golden age when Saturn ruled as 'King of heaven.' In applying the same language today, we would identify our sun as the current 'King of heaven.'

Hesiod and Ovid claimed[26] there had been a Golden Age during which Saturn (Kronos to the Greeks) had been the King of heaven, followed by a destructive flood and then a silver age during which Jupiter (Zeus to the Greeks) ruled as King of heaven, followed by the age of the Trojan war and then our present (Iron) age. Again, in the same language, our sun is the "King of Heaven" now . In today's language, this amounts to claims that Jupiter and Saturn had once been dwarf stars, that Earth was originally in orbit around Saturn, and then spent a shorter period of time orbiting Jupiter before our solar system settled into its present configuration. This postulated historical progression has become known in certain circles as "Saturn Theory"[27].

Over the past three to four decades there have been several versions of Saturn theories floating around; the only thing that we are fairly certain of is that one of these versions will eventually find general acceptance in the broader scientific and historical communities. There are versions that have Saturn approaching our present sun from vast cosmic distances, a version that has Saturn originally having fissioned from our present sun via an electrical process, and then there is a version that we favor. That version involves Saturn

---

[25] Plato "The Statesman", http://www.gutenberg.org/ebooks/1738
[26] Hesiod's "Works and Days" and Ovid's "Metamorphoses", e.g.
http://www.sacred-texts.com/cla/hesiod/works.htm "First of all the deathless gods who dwell on Olympus made a golden race of mortal men who lived in the time of Cronos when he was reigning in heaven..."
[27] http://www.maverickscience.com/saturn.htm

and our present sun having originally been formed in relatively close proximity by the same kind of z-pinch phenomenon we see displayed by cosmic Birkeland currents when creating groups of stars and strings of galaxies. This last possibility suggests a scenario entirely similar to imagining the dwarf star component (Proxima Centauri) of Alpha Centauri spiraling into one of the two main stars that constitute Alpha Centauri. We view this as the most likely scenario.

All of this is at extreme variance to what is taught in our schools. Our schools teach that the Solar System is several billion years old and has been in its present configuration for hundreds of millions of years. There is, however, a great deal of physical evidence to support the "Saturn Theory" alternative. Consider the axis tilts of the planets in our system. The problem from the perspective of the standard theory is obvious enough; if our system had formed from a swirling disk of solar material as claimed, all axial tilts should be approximately the same, that is, all near zero with all axes of the planets roughly perpendicular to the plane of our system. But that is not what we observe. Instead we observe great variations amongst the axial tilts of the planets, indicating that something is amiss in the accepted model of our Solar System's formation.

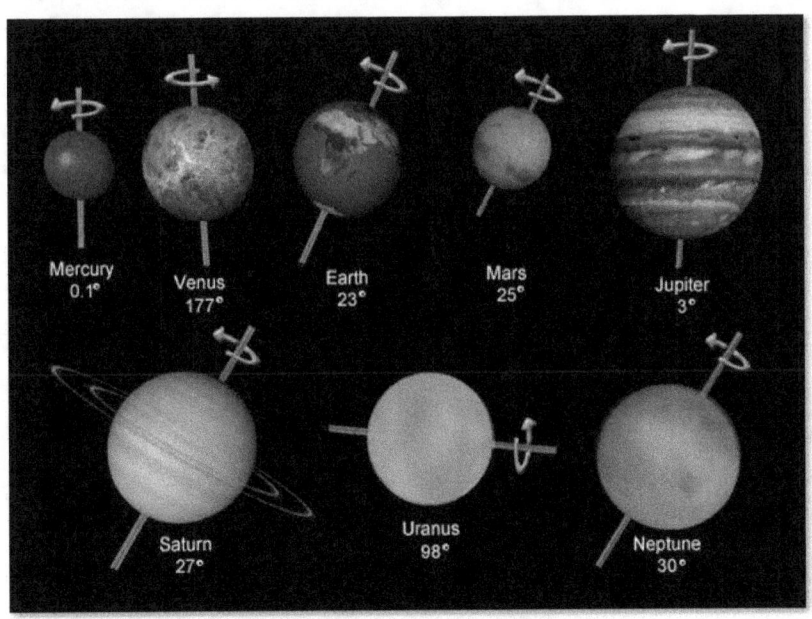

*The axial tilts of the eight recognized planets making up the Solar System. Of the eight, Earth, Mars, Saturn and Neptune share virtually the same degree of tilt, an indication that they are of the same family but foreign to our present Solar System, **i.e. that our present sun captured them as a group.** Mercury and Jupiter also share virtually identical near-zero-degree tilts (roughly perpendicular to the plane of the system), indicating their origins within the Solar System. The odd ones out are Uranus and Venus, the later identified in Saturn Theory as a late-comer violently ejected from Saturn during the chaotic height of Saturn's entry into the Solar System. The origins of Uranus' side on tilt is more problematic, though it is likely this gas-giant was also part of the Saturn family of planets, yet suffered a pole shift during that system's catastrophic capture by the Sun. Image not to scale.*

The highlighted item above is crucial to the narrative of this book and to Saturn theory in general. What is indicated is that a reasonable interpretation of those axis tilts entirely corresponds to the interpretation of myth and ancient literature and iconography which is associated with the Saturn theory. Again, Venus is problematical. The fact that its spin axes is roughly perpendicular to the systems plane indicates that it is either primordial to the sun/Jupiter/Mercury system or has its origins after the capture of the Saturnian system. The retrograde/backwards spin cannot be primordial and must have arisen via interaction with another body in the system. The curious phase lock with Earth (showing viewers on Earth the same face at inferior conjunctions) indicates that the other planet in question was Earth. The 900°F surface temperature, 90 bar CO_2 atmosphere, massive thermal imbalance, large upwards infrared flux, and numerous other features indicate that Venus is basically a new planet, which is in keeping with the interpretations of myth and iconography associated with the Saturn theory.

The list of anomalies within accepted historical and scientific paradigms extends beyond mainstream thinking on how our Solar System was supposedly formed. For example:

- We are told that modern man has been around for a couple of hundred thousand years and that recent

44

ancestors have been around for five or six million years, and yet we only have three or four thousand years of recorded history that we know anything about or about which anything could be said with any degree of certainty.

- We are told that humans share genetic components with Neanderthals; we are not told why there is no physical evidence on the planet of any cross-breeding between modern humans and Neanderthals[28] or any other hominid despite humans and Neanderthals having lived in close proximity for long periods of time in the Levant, where much cross-breeding would have been expected.

- We are told the dinosaurs perished tens of millions of years ago and yet scientists are now finding soft tissue including blood, blood vessels, collagen, and raw meat in dinosaur remains.

- We are told that dinosaurs dominated the earth for tens of millions of years and that their sizes were a winning ticket during that huge space of time; we are not told why, in the tens of millions of years that supposedly intervene between their age and ours, nothing else has re-evolved to those kinds of sizes.

- We are told that evolution is driven by a combination of mutation and natural selection; in the case of birds, we're not told what kind of mutation would change down feathers into flight feathers only on the creature's wings where they were needed.

- As noted already, it is known that the associations of the innermost six planets with the names of the ancient pantheon gods and goddesses are primordial; we are not told why the two chieftain gods of every one of those ancient religions were Jupiter and Saturn and not the sun and the moon.

- Unexplained also is the curious non-relation between Indo-European and Semitic languages[29];

---

[28] James Shreeve, "The Neanderthal Peace", http://discovermagazine.com/1995/sep/theneanderthalpe558#.UVHN1ztIZ 44

[29] http://web.cn.edu/kwheeler/IE_Non.html The IE/Semitic non relation is explored further in Appendix A

there is no meaningful racial difference between the two groups and they could not have split up more than a few thousand years ago.

The above points are food for thought in a world in which most are convinced that science has worked out all the major issues and is now simply ironing out details.

When Saturn Met the Sun and its Companion, Jupiter

The ancients speak of two distinct epochs for the world on which we live; the world we see today with the Sun and Moon as our prime celestial masters, and a time before that, when the heavens were ruled by the planet Saturn. This previous era is generically referred to in world mythology as the Golden Age, a time when Saturn shone forth as a sun in its own right from out of Earth's celestial north to bath the world in luxuriant and richly fertile warmth. This was Saturn at its zenith, the then chief star/god of Creation and Time, our first and "best" sun[30]. Vestiges of this ancient reality are not difficult to turn up. The chief religious festival of ancient Rome was called Saturnalia, and we still refer to the Sabbath as "Saturn's Day," or "Saturday".

But Saturn's Golden Age did not last.

As inconceivable as it may seem to the modern mind, the witness of ancient mythology reveals that mankind suffered a great cosmological upheaval in which Saturn's glorious rule was replaced first by Jupiter, and then eventually by the Sun. The consequences of this upheaval has left an indelible mark on the collective human psyche leading to a pervading doomsday anxiety, a fretful psychosis fed by deeply entrenched psychological archetypes that cross all known cultural barriers. In recent years, this pervasive fixation with large scale catastrophe has nearly become Hollywood's entire stock in trade. Deep down, whatever our cultural heritage, we all share in a seemingly irrational fear of impending catastrophe, a fear, born of

---

[30] For example: http://hans.wyrdweb.eu/tag/egypt/ Alchemists referred to Saturn as a "best sun" into the middle ages.

prehistoric experience, that has spawned the dominant themes in our religions, arts, architecture, and even our geopolitical and economic systems.

Yet, even before the coming of the Saturnian Golden Age, we hear of a time before time, a distantly primeval world calling to us from out of a universal dreamtime. These are the fading memories of humankind preserved in the oral traditions of our most ancient races and cultures. Through these dream-like fragments we hear echoes of an age spent in darkness, a twilight world existing long before the coming of the Sun. We hear of it as a time when the dull primordial half-light of the star/god Saturn shifted listlessly on the chaotic celestial waters that enveloped all human existence. Here we find the purple-hued dawn of creation, the time before time began; the time before bright, clear light entered into the world.

Through the work of comparative mythologists and, more recently, the contributions of plasma-based physicists, a new cosmology has emerged that seeks to explain these almost forgotten human legacies. This new cosmology identifies the mythical god Saturn as a primordial sub-brown dwarf star existing and drifting beyond the Solar System and accompanied by its own family of satellites. Our Earth was one of these satellites. Safely encapsulated within Saturn's protective and opaque bubble-like plasma sphere, the oral traditions tell us Earth's inhabitants enjoyed a warm, yet dully lit environment devoid of any concept of measured time and oblivious to the greater cosmos. In the swirling, enveloping chaos that was Saturn's outer plasma sphere nothing moved with regularity. Saturn's pale disk glowed from a fixed, if listless position above the Earth's North Pole, the position identified by world mythologies as the abode of the gods.

And slowly, yet surely, Saturn and its family of planets unknowingly drifted towards a fateful encounter with a much larger and more electrically dynamic star we now call the Sun. When contact was made between Saturn's own plasma sphere and the Sun's heliosphere, all hell broke loose! Powerful electrical forces surged and discharged causing Saturn to overload and flare nova-like. In what must have seemed an instant, humankind's previous twilight existence was spectacularly extinguished by a brilliant, bright light shining forth from out of the celestial north. 'Day One' in the epic of a new creation story had announced itself and humanity now confronted a deeply changed world brimming with visual access to the greater Cosmos. Life would never be the same as humankind entered into Saturn's universal Golden Age, an age that only ended when the Sun's overwhelming electrical influence conspired with the planet Jupiter to violently banish Saturn to the outer celestial realms. The great primordial Saturnian system of planets was torn asunder and Earth, for so long chained in axial alignment to Saturn, was freed from its bindings to acquire a new orbit around a new master, the Sun.

Such events, mythology tells us, took place roughly ten thousand years ago.

Yet, there are *two* conundrums we face in coming to grips with this radical interpretation of Earth's mythical and cosmological relationship to Saturn. Even if we accept that Saturn originally started off as a free-floating *sub*-brown dwarf star with an inhabited Earth as one of its satellites, we are faced with the mythological record's insistence that Saturn remained fixed and immobile at Earth's celestial north, in the same manner we see the Pole Star today. While most would offer a phase-lock solution to Saturn's perceived immobility, with Earth orbiting Saturn in the same way that the moon orbits Earth today, this does not explain the latter epoch in which Saturn remained seemingly stacked above Earth while both bodies circled the Sun on its equatorial plane. This forces us to consider Earth as having been in axial alignment with Saturn's southern pole during this time, an arrangement that was most likely a continuance of its existing primordial relationship to Saturn before the latter's catastrophic encounter with the Sun. However, this polar configuration, that of planets being stacked on top of one another in axial alignment, is viewed as an abomination by virtually all astrophysicists, an impossibility given the known tenets of celestial mechanics. This, then, was our *first* conundrum.

Saturn Theory Day at the Planetarium. The planets Mercury and Jupiter, the latter with its four Galilean moons, orbit the Sun as the axial aligned planets dominated by Saturn approach from below the Solar System at a roughly 27° angle to the Sun's own axis. The scene depicts events shortly before the Sun's electrical field provokes an electrical imbalance in Saturn causing it to flare nova-like. The planet Venus is absent at this time due to it not yet having been ejected by a flaring Saturn, an event that is preserved in mythology as the birth of the goddess Venus.

Then we come to the subject of *us, that is,* humankind's place in the general scheme laid out by Saturn Theory. More precisely, we are concerned with the assumed origin of human beings on an Earth in which ancient oral traditions tell us we had once stumbled around in near darkness for a period stretching back into unknown antiquity. This reported dim and inhospitable environment conflicts with the basic idea that a species is generally adapted to their environment of origin — *we*, human beings, are creatures fundamentally adapted to function best in a *bright* environment. This, then, was our *second* conundrum.

[Troy speaking here]

In discussions held between this work's two authors before the beginning of its writing, Ted found himself questioning the origins of human beings in a permanently and predominantly darkish world. That humans were present during this primordial epoch was undisputed — we see evidence of their remains ranging from great artworks painted onto the walls of deep caves to the scattered remnants of their stone tools around prehistoric camp sites — yet there was an uneasy realisation that something was not quite right with the scenario of humans having been originally born into this world of gloom.

In an exercise reminiscent of the classic Sesame Street children's game 'one of these things is not like the others,' Ted had pointed out that human eyes and eyesight would have been poorly adapted to this purple dawn-like environment. He demonstrated that, while other creatures, especially those of a more ancient origin, seem to display visual characteristics geared primarily to a nocturnal existence, humans find themselves at a fundamental disadvantage in a nocturnal environment without the aid of artificial light. Humans are one of those things that stick out in the darkened purple dawn scenario as 'not like the others'. They simply don't seem to belong. Interestingly, the other notable species with a similar disadvantageous trait is the dolphin.

The implications of Ted's observations only came to light, so as to speak, when the penny dropped as to what Ted was really saying. The inescapable conclusion was that humans did not necessarily originate on Earth! As a species, we seek the light. Our optimum environment is a bright environment. For humans to be trapped in a permanent gloom where one of our most important faculties is handicapped to the point of near uselessness vigorously militates against the concept of a permanently darkened primordial Earth being our environment of origin.

So we began to look elsewhere — and the journey we undertook eventually led us to a basic question that opened up a huge vista of potential solutions. The question that changed the game for both of us was the obvious flipside to Earth having begun life under a wandering sub-brown dwarf star called Saturn: *In the meantime, what was happening around the Sun, and was there a brightly lit world on which it was happening?*

Finding the answer to this question offered more than both of us anticipated. Ted eventually dubbed our subsequent findings to be a prehistory of what he called the Antique Solar System, a prearrangement of our current Solar System sans the planets Saturn, Neptune, Uranus, Mars, Venus and, of course, Earth. Subsequently, the role played by Jupiter during this Antique epoch seems to have contrasted sharply with Earth's experience under Saturn. Jupiter's moons also began to take on an increasingly important role in a story that, in one of those sweet ironies that only research of this kind can raise up, saw two avowed anti-evolutionists, Ted and myself, begin to entertain aspects in the theories of two hard-core evolutionists; Elaine Morgan's *Aquatic Ape Hypothesis* and Danny Vendramini's *Them+Us* theory of Neanderthal predation on ancient humans. (And if that last sentence involving Jupiter's moons, swimming humans and ravenous beastly Neanderthals doesn't pique your curiosity, then maybe nothing will.)

[Both authors speaking again]

Which brings us back to our *first* conundrum, namely that of explaining how a planet like Earth could find itself dangling down below a sub-brown dwarf star for most of its own distant prehistory? While at first glance this problem seems almost insurmountable from a purely physics point of view, a closer look at observable phenomena would suggest otherwise. By this we mean that there are, in fact, copious examples of axially aligned objects in our galaxy that offer clues not only to how axial alignments can function, but also as to the nature of how stars and planets are birthed in the first place. However, this is a discussion to take up in more depth in the next chapter, a chapter that will look closely at the role played by the astronomical phenomena called a Herbig-Haro object. Such objects, we will argue, establish the legitimacy of axial aligned celestial bodies in space and provide a solution to the seemingly bizarre relationship between Earth and Saturn in ancient times.

For those readers new to the above presented concepts, they could, at this point, simply choose to reject all cosmological notions of Saturnian planetary systems being captured in the distant past by the Sun and the resulting mythical cataclysmic orgies of destruction. They could simply dismiss the concept of an ancient and dark Saturnian primordial dawn as simply being the fundamental fallacy at the heart of our conundrum regarding human origins on Earth. They could offer the observation that daylight must surely have always been a feature in the human experience and that this is really why we have 'evolved' the way we are. Also, in seeking to reject the evidence presented by the mythological record for a dark primordial environment, they could lay charges that we have wilfully misinterpreted mythology to concoct our own flawed literalist dogma at odds with accepted scholastic and scientific consensus. To answer such critics, all we can say is read on if you dare and make your criticisms after absorbing the full argument presented in this work. We believe there is much to offer, if only for the purpose of constructive debate on topics that should intrigue anyone searching for an understanding as to where we come from and who we are.

Summary and Takeaways from this Chapter

The ancient world included broad categories of things that seem very strange to modern ears: planetary scale catastrophes, religious practices and rituals that amounted to using the human mind and brain in ways that are no longer possible, and astral religions in which the two chieftain gods were invariably Jupiter and Saturn, and not the sun and moon as would be the case were people to create an astral religion today.

Greco Roman literature describes world ages in which first Saturn and then Jupiter are called the "King of Heaven." In the same language, our sun would be the "King of Heaven" today. The ancients appear to have been expressing a belief that Saturn was recently the "sun" i.e. that it was very recently a dwarf star and that our own planet was in orbit around it.

Our sun, Jupiter, and Mercury have spin axes nearly perpendicular to the plane of our system, suggesting that those bodies formed an original system. Saturn, Mars, earth, and Neptune all have spin axes of 23 to 27° offset from being perpendicular to the plane of the system. This strongly suggests that these bodies were captured as a group by our present sun via some catastrophic event or series of events.

Ancient literature including Genesis as well as Greco Roman literature speaks of the Golden age prior to the flood at the time of Noah. Older traditions however, mostly oral traditions retained by the oldest human groups, speak of a "Purple Dawn" age prior to the Golden age, before there was any major source of light such as we experience now in our world.

Mankind's Purple Dawn

Dancing in the Dark

"In the beginning there was only darkness. Yet, in that darkness, there was already Raven. He was *still small and weak* and his special powers had not fully developed."[31] *Eskimo creation myth*

Between 20,000 - 40,000 years ago, humans in Europe entered into the dark and foreboding gloom of various deep cave systems, lit their animal fat fuelled lamps, and proceeded to produce artistic masterpieces that, when rediscovered millennia later, led Pablo Picasso to declare modern art "had invented nothing!" According to the scenario that will unfold in the following pages, the subterranean gloom encountered by these early artists was nothing substantially different from their everyday existence on the surface above. Outside these caves, in the open air of a world existing under a chaotic void-like sky, mankind's early ancestors are reported in myth to have lived in a perpetual twilight devoid of our current sun. Their world, according to many ancient creation accounts, was permeated by a dull glow that provided barely enough light by which to read your average Palaeolithic newspaper.

Dwardu Cardona, in a chapter in his book *God Star* titled 'The Age of Darkness,' points to comments made by P. Wheeler on the Japanese creation myth that indicate the universal nature of the Kronos/Saturn myth as the main construct in this primordial age of darkness.

---

[31] M. Wood, "*Heroes and Hunters from North American Indian Mythology,*" (N.Y. 1982), page 17; as quoted with emphasis added by Dwardu Cardona, "God Star," (2006), page 284. Note: 'Raven' is the Eskimo equivalent to the primordial star and later planet Saturn.

"In the earliest legend with which the recital [i.e., the *Kojiki*] opens, one recognises the primal myth . . . the development from a primordial darkness and chaos. . . This is the Kronos legend, in its thousand forms, the father of all mythologies, upon which so many peoples have constructed their cosmogonies."[32]

While also providing many direct quotes from various world creation myths, including Jewish literature, Cardona cites H. Osborne whose work on South American mythology also recognised this primeval 'age of darkness' theme:

"Some mythological cycles feature a primitive age of darkness *before the existence of the sun,* when human beings lived in a state of anarchy without the techniques of civilized life."[33]

An age *before the existence of the sun?* As noted by Cardona, such descriptions may leave the reader with the mistaken impression that, without the sun, there was no light at all. Yet, this is not quite the case in mythology, and the same creation myths point to the existence of at least a modest amount of light before the coming of the sun, albeit from a different source to our current sun. That source was Saturn, the ancient creator god in his myriad of forms throughout world mythology, and conclusively identified with the planet Saturn. This celestial Saturn was the very same *small and weak* Raven-character we saw in the Eskimo quote at the beginning of this chapter.

---

[32] P. Wheeler, *"The Sacred Scriptures of the Japanese,"* (N.Y. 1952), page 387; as quoted by Dwardu Cardona, "God Star," (2006), page 275.

[33] H. Osborne, "South American Mythology," Mythology of the Americas (London, 1970), page 294; as quoted with emphasis by Dwardu Cardona, "God Star," (2006), page 278.

The Purple Dawn of Mankind. Artist's impression of the mythical 'age of darkness' when Saturn is said to have hovered as a pale, weak orb in Earth's north celestial sphere. The purple hue of the light is a result of Saturn's blue/red light spectrum, the spectrum most associated with brown dwarf stars.

Small and Weak

The light of our current sun, as a rule, is *not* essential to life. Microbial life can exist in a sunless environment as can species of deep sea life, while photosynthesis in vegetation works best in the darker, red-light spectrum rather than the harsh and bright ultraviolet light of the Sun. So it is not inconceivable that life on Earth as we know it could have existed, and even flourished, in a predominantly nocturnal world, provided there was some form of radiated red-spectrum light or energy. In fact, most species of animal, including those now long extinct, exhibit high degrees of nocturnal adaptation. Of all the higher creatures currently inhabiting the Earth, human beings are probably the least adapted to a nocturnal environment.

Yet traditional mythologies remember a time of darkness stretching out into an unknown antiquity, a time in which the god Saturn, in all his manifestations, was small and weak, a mere shadow of the creative force he was destined to become. As mankind's first remembered source of light, long before the coming of the sun, Saturn is said to have cast its pale light on a world without seasons and devoid of any means for humans to calculate time. Locked in a stationary post that is reported in mythology to have been at the northern celestial realms, the primordial Saturn seemingly drifted aimlessly through the skies on a chaotic heavenly milieu resembling the ebb and flow of a dark ocean. It is this state of affairs that is referred to in the opening verses of Genesis when "darkness [was] upon the face of the deep. And the spirit of God moved upon the face of the waters."[34]

Our conclusion here is that Saturn, this primordial source of a dim, timeless light, was a dully glowing *sub*-brown dwarf star of which Earth was one of its primary and original satellites; and that Earth was nestled close enough to its original host star to have been enveloped in Saturn's opaque and warming plasma sheath. For humans alive on Earth at this time, there would have been no reference to the greater cosmos, and therefore no reference to any moving celestial object with which they could have marked the passage of time. This was a timeless age in which the dim blue/red-light spectrum emanating from Saturn would have cast a dark purple-hued glow over Earth's surface. This, then, was mankind's purple dawn of creation.

Life Under a Brown dwarf Star dwarf

Electric Universe (EU) physicist Wallace Thornhill has suggested that planets orbiting closely to brown dwarf stars would be the best place to go looking for life as we know it outside the Solar System. This is a possibility under the EU model because all types or stars, including brown dwarfs, are explained as an electric discharge phenomenon taking place where vast cosmic and electrically live Birkeland currents entwine and pinch down into what is called a z-pinch (also known as a Bennett pinch). Discovered over one

---

[34] Exert from Genesis 1: 1-2.

hundred years ago, Birkeland currents,[35] or the filamentary gas-like strings of twisting 'plasma ropes' seen in space, are viewed in the EU model as the galaxy's power lines feeding electrical power to all the stars we see shining in the night sky.[36]

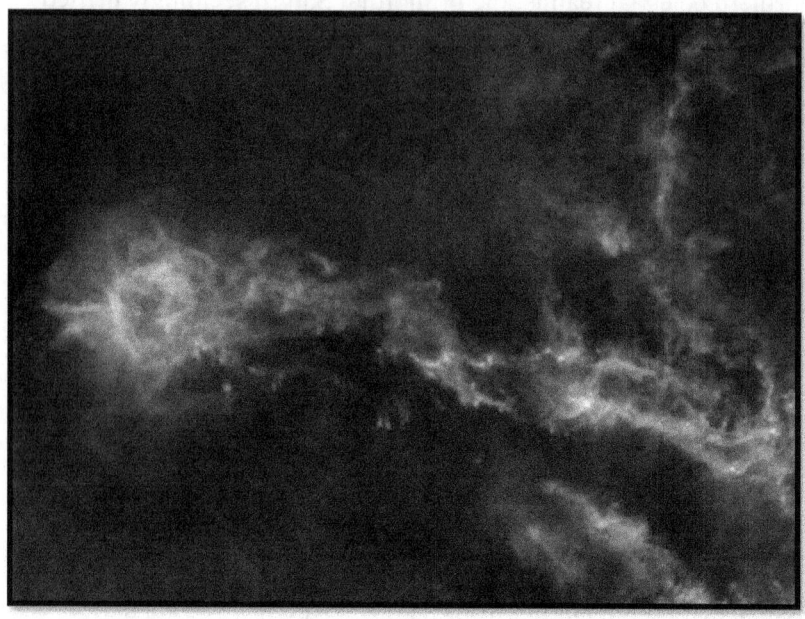

Mistakenly referred to as filaments of 'hot gases' that are subject to meteorological-like space winds and mechanical models producing

---

[35] For an overview on Birkeland currents, see:
http://en.wikipedia.org/wiki/Birkeland_currents
For an Electric Universe take on the role played by Birkeland currents in the formation of galaxies, see: Wallace Thronhill, "Electric Galaxies," May 20, 2008, http://www.holoscience.com/wp/electric-galaxies/

[36] An excellent summary of Birkeland currents acting as galactic power lines can be seen here:
http://www.thunderbolts.info/tpod/2011/arch11/110629powerlines.htm
Also see: "Cygnus Loop," *Picture of the Day*, August 31, 2005, Thunderbolts.info,
http://www.thunderbolts.info/tpod/2005/arch05/050831cygnusloop.htm

acoustic shocks,[37] the filamentary orange-like strings seen in this image are actually made of plasma, an excellent conductor of electricity. These are electrically alive Birkeland currents operating at interstellar scales, seen here feeding electrical energy into a star manufacturing area called the Cocoon Nebula (blue region). Image credit: ESA/Herschel/SPIRE/PACS/D. Arzoumanian (CEA Saclay)

The z-pinch effects we see in Birkeland currents appear as beads along the giant strings of interstellar plasma seen in many telescope images taken of deep space. These glowing beads, like pearls on a string, are the points of light that make up the huge clusters of stars seen in ours and other galaxies.[38] The Birkeland currents feeding these stars or beads with electrical power are usually in their dark mode and therefore cannot be detected by traditional optical means.

> "As the effect, called a "z-pinch," increases, the electric field intensifies, further increasing the z-pinch. The compressed blobs form spinning electrical discharges. At first they glow as dim red dwarfs, then blazing yellow stars, and finally they might become brilliant ultraviolet arcs, driven by the electric currents that generated them."[39]

---

[37] (Note to photo caption) For an Electric Universe perspective on the debate surrounding gaseous filaments seen in space, see: "Hot Gas vs. Electric Currents," *Picture of the Day*, April 17, 2009, Thunderbolts.info; http://www.thunderbolts.info/tpod/2009/arch09/090417hotgas.htm

[38] See: "Electrical Birthing of Stars," *Picture of the Day*, March 4, 2005, Thunderbolts.info; http://www.thunderbolts.info/tpod/2005/arch05/050304starbirth.htm

[39] Stephen Smith, "How Stars are Born," *Picture of the Day*, November 6, 2009, Thunderbolts.info

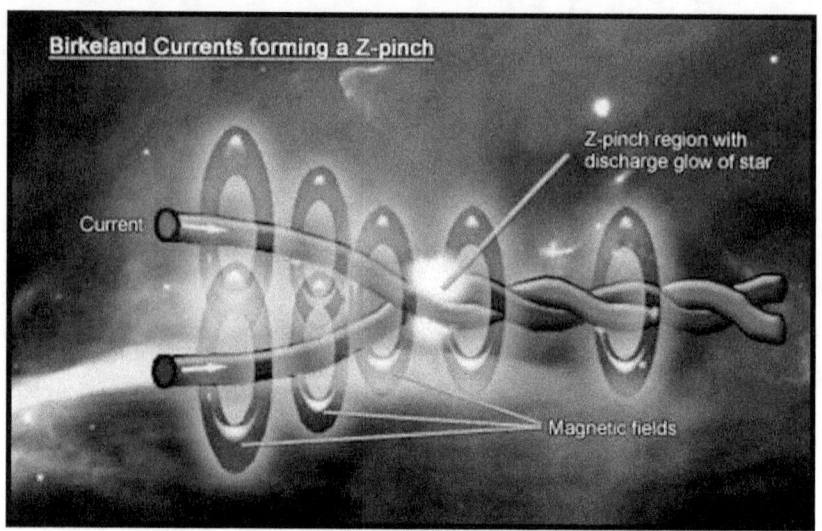

The vast clouds of filamentary plasma seen in photos taken of deep space cannot be detected by the human eye. This is because the plasma seen in them is operating in what is called 'dark mode.' Only when plasma shifts into 'glow mode' as seen in aurora phenomenon and 'arc mode' as seen in welding torches and stellar flares can it be seen by the naked eye. In the above graphic a simplified diagram of a Birkeland current z-pinching down to create a star discharge has been superimposed onto a classic false-color image taken of deep space by NASA.

The star's visible outward glow is the electrical anode of these z-pinches and it is analogous to the same discharge glow we see given off by an ordinary electric light bulb. As the late electrical engineer and scientist Ralph E. Juergens has said regarding our own sun, as quoted by Thornhill:

> "As I pursued the phenomenology of electric discharges, it gradually dawned on me that, structurally, the atmosphere of the sun bears a striking resemblance to the low-pressure type of electric discharge known as the glow discharge…"[40]

---

[40] See "Twinkle, twinkle electric star," by Wallace Thornhill, July 1, 2008; http://www.holoscience.com/wp/twinkle-twinkle-electric-star/

What this means is that if the temperature of a star's glow discharge is low enough, such as that of a brown dwarf star, then planets orbiting in close proximity would not experience the searing heat associated with main sequence stars like our sun.

> "Since an electric star [as opposed to a nuclear fusion-driven star] is heated externally [by Birkeland currents] a planet need not be destroyed by orbiting beneath its anode glow. In fact life is not only possible inside the glow of a small brown dwarf, it seems far more likely than on a planet orbiting outside a star! This is because the radiant energy arriving on a planet orbiting inside a glowing sphere is evenly distributed over the entire surface of the planet. There are no seasons, no tropics and no ice-caps. A planet does not have to rotate, its axis can point in any direction and its orbit can be eccentric."[41]

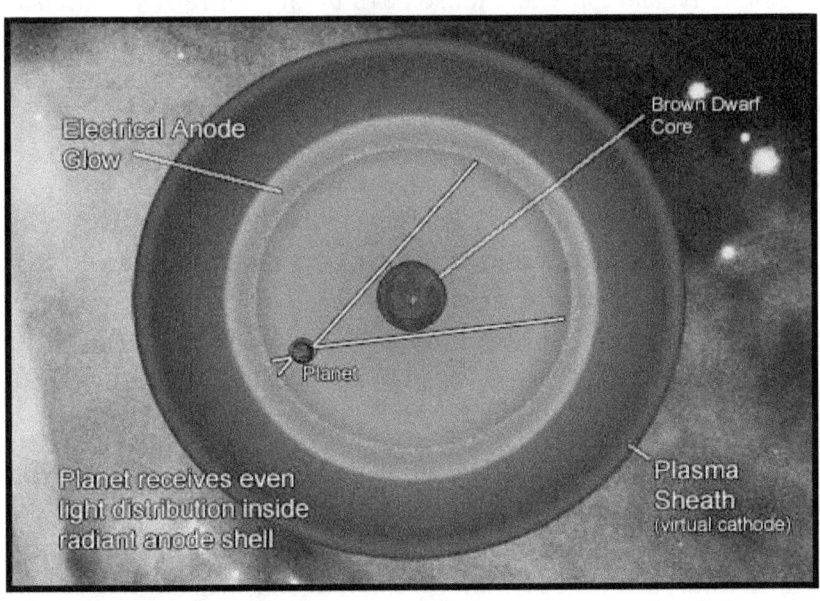

Brown Dwarf with planet existing inside its anode glow shell and plasma sheath according to the Electric Universe model for the

[41] Wallace Thornhill, "Other stars, other worlds, other life?" December 15, 1999. See; http://www.holoscience.com/wp/other-stars-other-worlds-other-life/

Here we have the type of conditions required for mythology's purple dawn of creation; i.e. a terrestrial-type planet trapped within a low temperature opaque glowing sheath and receiving enough radiated energy for life to survive, yet that life having no frame of reference beyond the all-encompassing dull anode glow of its host star. Brown dwarf stars generally glow in the 800 - 1,700 Kelvin range, a temperature range that can provide enough warmth to a close orbiting planet for human life to flourish. By contrast, self-regulating main sequence stars like our own sun glow at temperatures of 5,500° Celsius plus on their surfaces and in excess of 2 million Kelvin at their coronas; far too hot for a similar scenario involving any close orbiting planets with life on them.

Importantly, and especially for some of the Jupiter-related concepts that appear later in this work, Wallace Thornhill has also provided us with an illustration of just how such a planetary arrangement might exist inside the anode glow of a brown dwarf:

> "For example, consider Jupiter as an independent body moving in the galaxy inside its radiant plasma sheath (analogous to a cometary coma). It would be regarded as a brown dwarf star! And even if that glowing sphere were half the size of Jupiter's present magnetosphere, which is 10,300,000 km in diameter, all of Jupiter's large moons would orbit comfortably inside that cocoon."[42]

That the earth was originally encased in a similar situation under a brown dwarf star called Saturn is central to understanding why mythology tells us our most primordial existence was one of darkness, an age when Saturn hovered as a dull orb in the chaotic swirls of the northern skies. Outside stars could not have been seen

---

[42] Wallace Thornhill, "Assembling the Solar System," October 23, 2008. See; http://www.holoscience.com/wp/assembling-the-solar-system/

because Saturn's opaque plasma sheath and its anode glow would have blocked out all incoming light — much in the same way that the glow from city lights bouncing off the atmosphere can block out the stars today.

And herein lays a caveat in contemplating life under a brown dwarf star, for it was this enveloping plasma sheath that catastrophically short-circuited and lost its opaqueness when it eventually brushed against the Sun's heliosphere during Saturn's approach towards the Solar System, an event that is recorded in mythology as the dramatic and destructive flaring of the god Saturn at the start of the fabled Golden Age.

> ". . . The brown dwarf 'Garden of Eden' comes with a caveat. Stars off the main sequence do not have the self-regulating photospheric discharge to smooth out variations in electrical power input. Consequently, brown dwarfs are subject to sudden outbursts, or 'flaring,' when they encounter a surge in the circuit that powers them. These flares could cause sparking to and between the satellites orbiting inside the sheath and lead to sudden extinction events, vast fallout deposits and fossilization. There is much food for new thoughts!"[43]

Food for new thoughts indeed! We can only begin to imagine the effect on Earth's existing human inhabitants in seeing their previously passive host star burst into life. What terror must have been felt as their previously dark and tepid world suddenly disintegrated in a blindingly bright flash of light and the huge vastness of space was revealed to them!

Yet, what was this world truly like for those humans living there before these spectacular events that heralded the start of the mythical Golden Age? Can we truly ever come to understand the challenges and complexities of their lives as they struggled to survive and make meaning of the world in which they lived? After all, these are the same people who produced the masterpieces seen

---

[43] Wallace Thornhill, "Twinkle, twinkle electric star," July 1, 2008, see; http://www.holoscience.com/wp/twinkle-twinkle-electric-star/

in the caves at Lascaux and Chauvet and other parts of the world, a people whose artistic merits cannot be questioned and whose obvious knowledge of self cannot be denied. And in the same way that our modern interest in them belies a fascination with our origins as a species, could it also be that they too wondered from whence they had come, and who they might have truly once been?

The Mystery of Earth Under Saturn

"The evidence of myth which points to Saturn having once occupied a position above Earth's north polar regions is voluminous. There is not a race on Earth that has not preserved at least one account which states as much. According to this evidence, Saturn occupied a central position in the north celestial regions. It rotated, and rotated widely; but other than that, it was immovable."

Dwardu Cardona (1978)

A macro-cosmological view of the world in which our cave painting ancestors lived must take into account the mythological record of a dominant Saturn, the very same god that was clearly identified by the ancients with the actual planet Saturn, a planet that now, somewhat incongruously, resides in exile in the outer realms of the Solar System.[44] As noted already, this reportedly stationary celestial object hovering over Earth's northern polar regions eventually burst forth into a bright sun in its own right where it claimed the mantle as Earth's original sun. E.S. Butterworth had this to say about how the ancient's viewed their sun:

> "[The sun of the ancients] is not the natural sun of heaven, for it neither rises nor sets, but is, as it seems, ever in the zenith above the navel of the world. There are signs of an ambiguity between the pole star and the sun."[45]

---

[44] While critics have tried to refute a direct association between the mythical god Saturn and the planet Saturn based on claims that the naming of the planets only occurred at a later date and are therefore coincidental with mythology, this can clearly be shown not to be the case. Such spurious refutations stem from the works of a 19th century scholar called John J. O'Neill who, in his work *The Night of the Gods*, recognised the consistent placing of Saturn above the north pole, but contrived to create a distinction between the god and the planet due to his inability to reconcile the mythological record with the planet Saturn's current orbit. See: David Talbott, "Guidelines to the Saturn Myth." *KRONOS X:3* (Summer 1985)

[45] E. A. S. Butterworth, "The Tree at the Navel of the Earth," (Berlin, 1970), page 124

The *Popol Vuh*, a Mesoamerican text detailing similar traditions, tells of the same phenomenon associated with the Saturn myth:

> "Like a man was the sun when it showed itself. It showed itself when it was born and *remained fixed in the sky* like a mirror. *Certainly it was not the same sun which we see*, it is said in their old tales." [emphasis ours]

There are only two ways a celestial stellar object can appear to remain immobile in the sky from the perspective of one viewing it from a rotating Earth. The first is if the Earth were in phase-lock with the object in the same way the moon is in phase-lock with the Earth today. Or, and this is the more controversial solution, the Earth is suspended below Saturn's south pole where it rotates in axial alignment with its host star (see graphic below). While variations of the former configuration, the phase-lock solution, are favored by the majority of Saturn Theory researchers, the great champion for the latter solution is Dwardu Cardona. Dubbing his model the *polar configuration*, Cardona tells of the difficulties this solution faces while giving the reasoning for his acceptance of such a model:

> "[Speaking of Prof. Lynn Rose's phase lock solution] It is, therefore, understandable that, in *his* Saturnian scenario . . . Lynn Rose opted for an Earth in phase lock with Saturn. And yet for years I had reason to object to Rose's explanation of this phenomenon, just as he found reason to object to mine — since, almost from the very start of my research, I had come to the conclusion that Saturn's proper placement in Earth's primordial sky *had* to have been in Earth's north celestial sphere. Given that Rose's model is more feasible from a physical point of view, why do I opt for this bizarre idea? The answer is simple enough: *That* is where the mytho-historical record places the primeval Saturn — plumb in the centre of the Earth's north celestial

sphere, the very place which is presently occupied by the Pole Star."[46]

Cardona's point coincides with the question of Egyptian art forms (reproduced in chapter "Splash Saltations") showing a rotation about a central northern pole, i.e. the ship-of-morning, ship-of-day, ship- of-evening, ship-of-night images.

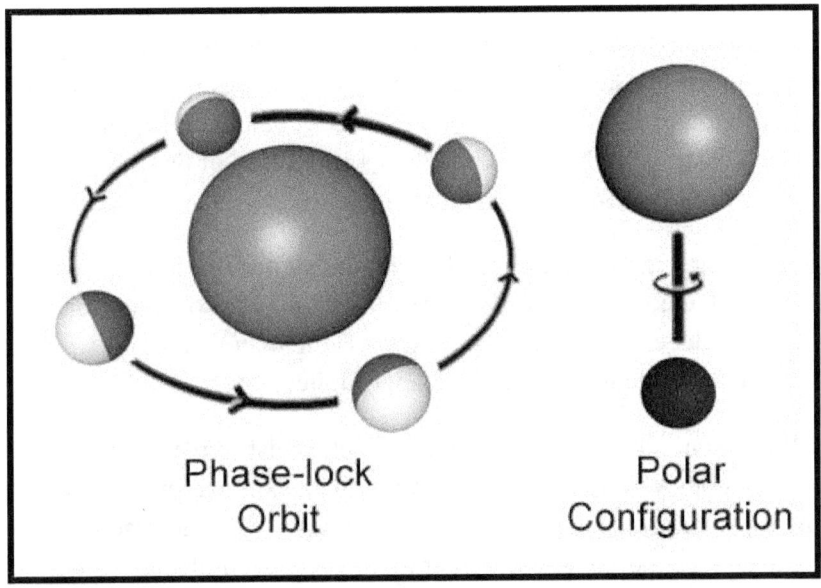

The two competing models describing how Saturn could have been seen by inhabitants on Earth sitting immobile at Earth's celestial north.

Readers with a physics mind-set, especially those with a basic grounding in Newtonian celestial mechanics, will find the concept of planets aligned axially in a polar configuration untenable. This is

[46] Dwardu Cardona, "God Star," *Trafford Publishing*, Victoria B.C. Canada, (2006), pages 220 – 222.

especially so since it seems obvious no such configuration is observed amongst any of the planets currently known. However, there have been observations of such a phenomenon, just not with planets — yet! In fact, polar configurations in their various guises are becoming increasingly common in the observable universe, providing us with vital clues into the physics of such a model. It's simply a matter of knowing what you are looking at.

Broken Comets and Star Factories

Something extraordinary happened in 2005 — contrary to all previous known laws of physics, scientists at the California Institute of Technology confirmed that bumble bees could, and actually do indeed fly.[47] And the world breathed a sigh of relief that all was well with the world. . .

Something extraordinary had also previously occurred in the year 1994 — the comet Shoemaker-Levy 9 split apart into twenty-one separate pieces, reformed into a *polar configuration*, and spectacularly slammed into the planet Jupiter.

Of the event taking place in 1994, astrophysicists have been conspicuously quite in coming forth with an explanation as to why the separate pieces of comet Shoemaker-Levy 9 should assume its famous 'string of pearls' configuration where each piece was stacked up above the one below it in axial alignment.[48] To date most physicists are only concerned with the fireworks surrounding the impact of Shoemaker-Levy 9's pieces as they crashed into Jupiter where, contrary to all mainstream expectations, they exploded in the higher Jovian atmosphere and not further down towards the planet's denser interior — but that is another story perfectly explainable by the EU model, which sees such 'impacts'

---

[47] See: "Deciphering the Mystery of Bee Flight," Pasadena, California, *NEWS*, 11/29/2005, see: http://www.caltech.edu/content/deciphering-mystery-bee-flight

[48] An excellent diagram detailing Shoemaker-Levy 9's polar configuration, reproduced in part in the following page's graphic can be seen here: http://ase.tufts.edu/cosmos/view_picture.asp?id=1262

as mostly attributable to destructive atmospheric electrical discharges.

For the issue at hand, the fact that Shoemaker-Levy 9's shattered remnants formed up into precisely the type of celestial alignment posited for the Earth/Saturn polar configuration as recorded in ancient times should give critics pause for thought in their assertions that no such celestial configuration can occur. Here is observed evidence for just such a configuration, albeit at a vastly smaller scale.

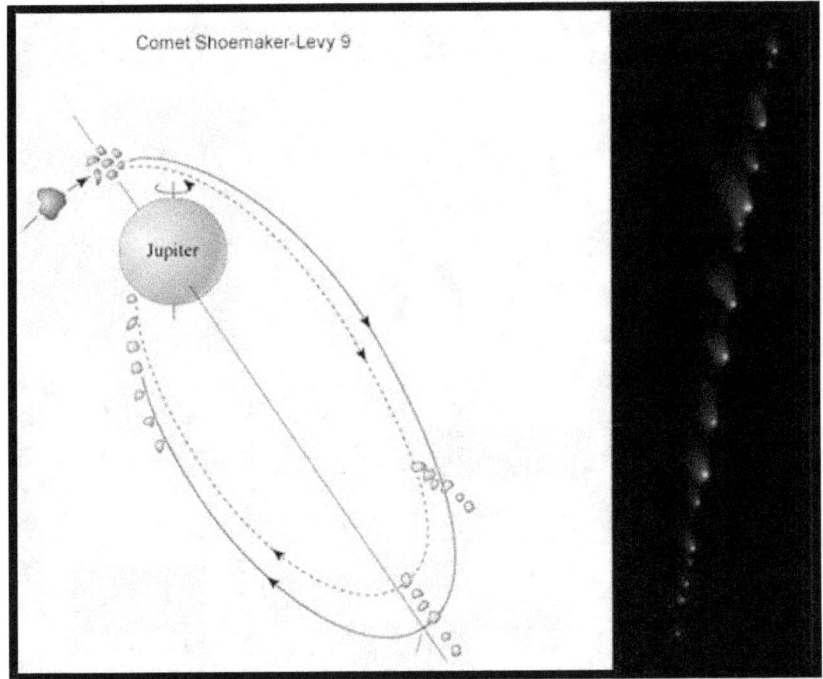

Diagram showing comet Shoemaker-Levy 9's breakup and subsequent orbit in its famous 'string of pearls' formation – i.e. each piece axially aligned. At right is a photo showing all twenty-one pieces of the broken apart comet. Diagram adapted from original image by: Professor Kenneth R. Lang, Tufts University; Photo image credit at right: NASA, ESA, and H. Weaver and E. Smith (STScI)

Then again, at the other end of the cosmic scale we have an increasing number of observed axially aligned space objects called Herbig-Haro objects. According to mainstream science, "Herbig-Haro objects are ubiquitous in star-forming regions, and several are often seen around a single star, *aligned along its rotational axis.*"[49] Herbig-Haro objects are recognized space regions viewed as star-making factories.

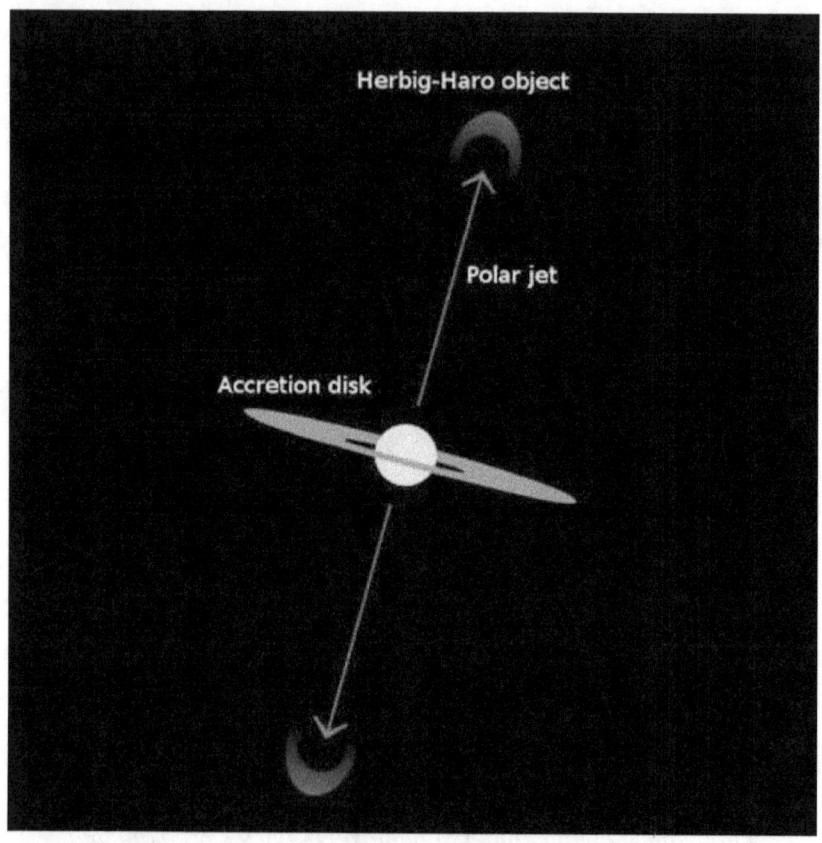

The standard (and badly labeled) model for a Herbig-Haro object showing the axially aligned 'polar jets' being expelled by a newly formed proto-star. The half-moon objects at the ends of the polar jets are thought by mainstream science to be 'bow shocks' caused by the fast moving hot gas of the jets moving through cold interstellar space. The disk labeled an 'accretion disk' is the circumstellar cloud of dust and debris found around most stars — we reject the

---

[49] Wikipedia entry on "Herbig-Haro object," see: http://en.wikipedia.org/wiki/Herbig%E2%80%93Haro_object

notion that any form of gravity-only induced 'accretion' is taking place here.

Herbig-Haro objects are believed to be associated with *proto*-stars in their infancy, and the accepted view is that these baby stars are shooting out vast polar jets of gas along their rotational axis in which globules or beads of plasma collect in our now familiar 'string of pearls' analogy. These so-called beads maintain their axial alignment and rotation in step with the proto-star exactly in the manner suggested for the Earth/Saturn polar configuration. Here, in fact, is the elusive evidence pointing to the possibility that polar configurations are possible in the depths of space.

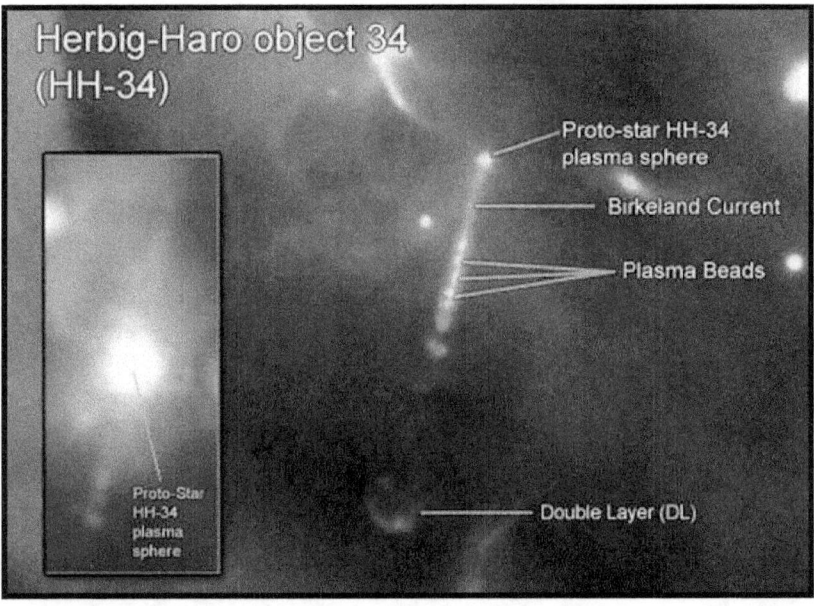

The Herbig-Haro object 34 (HH-34) is a classic example of axially aligned celestial objects conforming to the polar configuration model postulated for an ancient Earth/Saturn axial alignment. The inset photo is a Spitzer Space Telescope image with enough resolution to show HH-34's northern polar jet/Birkeland current. Image credits: NASA, IRAC

However, astronomers are not quite sure what is going on inside these beads of plasma. There is not enough infrared radiation to convince them that more main-sequence proto-stars are being born and they have certainly never suggested that the formation of proto-planets or sub-stellar objects might be taking place. We, on the other hand, claim this may be exactly what is happening.[50]

Why?

Firstly, what mainstream scientists identify as polar jets of hot gas are, in fact, Birkeland currents, the great interstellar and interplanetary transmission lines for the flow of electrical energy through interstellar space. The beads of plasma collecting along these Birkeland currents are where z-pinches are taking place. Z-pinches are extremely stable areas into which heavy elements like iron, ejected from the proto-star or drifting in interstellar space, are attracted and captured due to the intense magnetic fields associated with z-pinches. Some of these z-pinches fail to spark into full-blown main-sequence stars, and instead produce brown dwarfs, They even may produce the solid cores needed for the formation of terrestrial-type planets — and all this is happening along the same axial alignment of their proto-star's shared rotation.

Secondly, the so-called 'bow-wave shocks' supposedly produced by the hot gas shooting out along the proto-star's polar axis are nothing more than what plasma physics calls a 'Double Layer'. These double layers, or DLs, are the signature effect of a Langmuir sheath, or, in other words, a *plasma sheath*; the same protective electrical cocoon we have already encountered when looking at the electrical environment surrounding brown dwarf stars. Their presence in Herbig-Haro objects is a dead giveaway that serious electrical activity is taking place, the kind of activity that produces intensely strong and attractive magnetic fields. Wherever you have powerful

---

[50] Recent work by the Gemini Observatory has determined the presence of iron in similar 'gas bullets' being shot out of the Orion Nebula. Iron is, of course, a precursor element to the forming of a core around which a planet or a star can form, according to the EU model. See: "Gemini's Laser Vision Reveals Striking New Details in Orion Nebula," Gemini Observatory, Hilo HI, USA, http://www.gemini.edu/node/226

magnetic fields you have a recipe for potential planet and star-birthing activity; it is the magnetism at work that attracts heavy elements like iron to form a solid core. It is this profusion of electrical activity that is most relevant to our assertion that Earth started off under Saturn according to this polar configuration model.

Saturn's Birth and the Birth of Planet Earth

Obviously, the mainstream nebular hypothesis for planet formation out of a star's so-called accretion disk is discarded where this work is concerned. There are too many problems with this hypothesis, mostly to do with mainstream astrophysics' insistence that gravity, and not electromagnetism is the dominant force involved in shaping star systems and the creation of their planetary satellites. When electric forces are factored in, it is the power of a Birkeland current's z-pinch that determines how and where heavy elements are accreted (magnetically attracted) to form the core for any star, planet or sub-stellar object. Herbig-Haro objects proliferate with these powerful yet stable magnetic fields making them ideal candidates for the birthing of brown dwarfs and planets.

The sheer size of some of these axial aligned Herbig-Haro objects suggests a vast degree of separation between the main proto-star itself and the beads of plasma seen forming along its length and at the furthest extremities of the proto-star's electrically powered polar jets. However, should the Birkeland current emanating from the main proto-star electrically surge, then any star or planet-producing activity taking place in one of these beads would be severely impacted. If one of these beads harbors a proto-brown dwarf, then it is likely this proto-brown dwarf will flare electrically under the stress of the surge and eject a portion of its own core out along its own polar axis. This happens when an electrical short circuit takes place and the proto-brown dwarf's internal core fractures to leave two pieces of positively charged iron core — the smaller piece of the core is repelled (ejected) by the like-charged parent core.

All the above can potentially take place inside any one of these giant beads of plasma seen strung out along the length of most Herbig-Haro objects. According to this scenario, the plasma beads seen by our telescopes constitute the proto-plasma sheath for any newborn brown dwarf and any newly ejected moon or planet that is being held in place by a z-pinch in the proto-star's Birkeland current. This is why there is a low amount, if any infrared activity detected from inside these beads; it is not a main sequence star like our sun that is being formed there, but the conditions for the forming of a brown dwarf star or planet.

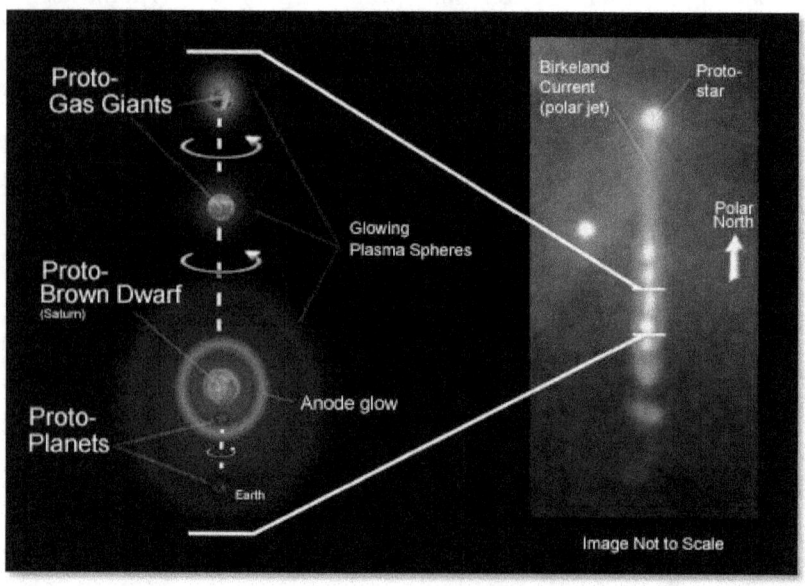

The prospective axial alignment of the Saturnian system of planets within the beaded plasma spheres found along a proto-star's polar jets. This diagram uses the axially aligned Herbig-Haro object 34 (HH-34) as an analogous reference. Image credit for HH-34 photo: NASA.

A brown dwarf star formed in this environment can be expected to maintain its own axially aligned Birkeland current even if it is severed from the main Birkeland current emanating from the main proto-star. Acting as a spinning homo-polar electric motor, or Faraday motor, the new brown dwarf star will generate its own electrical equilibrium as it feeds from the same general galactic electrical circuit that is also driving the main proto-star at the heart of the now breaking-apart Herbig-Haro object. In this way its axial tilt may change slightly, an important consideration when contemplating why Saturn came to have a *different* axial tilt to the Sun.

Any close proximity proto-planets captured along the Birkeland current of this newly formed and now separated brown dwarf star, either through ejection from a parent body or through the magnetic attraction of heavy elements into a z-pinch, can also be expected to remain trapped in the z-pinch in which they find themselves. While the vast majority of proto-planets and moons will eventually scatter like buckshot to find orbits along the equatorial plane of our newly formed brown dwarf, some will remain trapped in the z-pinches of the existing Birkeland currents flowing along the new system's rotational axis. Again, it is well to remember that z-pinches are very stable electrical constructs and are therefore quite capable of holding a planetary body in rotational lock-step alignment with the polarity of the brown dwarf's Birkeland current.

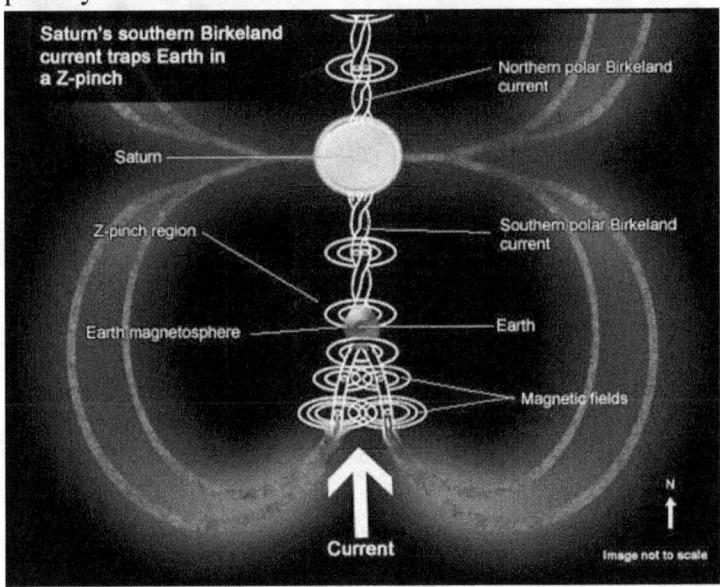

Birkeland currents forming Saturn's electrical circuit with Earth

75

trapped in a z-pinch along Saturn's southern polar Birkeland current. Saturn is operating as a giant Faraday motor that produces the powerful Birkeland currents seen here. The formation of an intrinsic magnetosphere around the Earth is a key component for allowing terrestrial life to exist in the high radiation field produced by a brown dwarf star like proto-Saturn. Graphic adapted from Hannes Alfvén's larger model for a galactic electrical circuit.

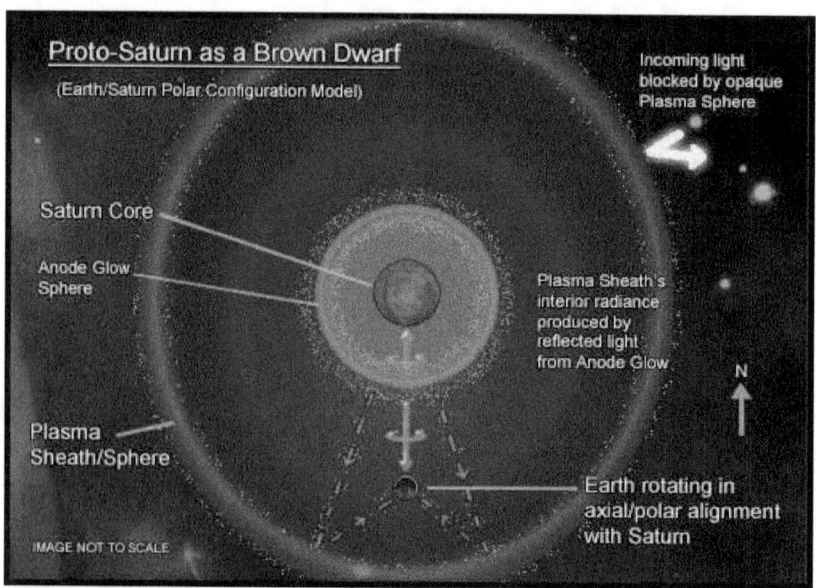

A cross section of the polar configuration model for a brown dwarf (proto-Saturn) harboring an axially aligned planet (Earth) between its anode glow discharge and its outer plasma sheath. Light from Saturn's pale anode glow is reflected back off its plasma sheath to provide a generally uniform spread of energy over the Earth at all points. Only Earth's northern polar region would receive direct light and would therefore be the brightest area on a fairly dark planet.

It should also be remembered that the lifespan of a Herbig-Haro object is a relatively short-lived affair, lasting in the tens of thousands of years, and not the millions of years usually associated with star formation. Things happen quickly where these objects are concerned, and they apparently begin to break up once the proto-star at its center develops into a fully-fledged main-sequence star like our current sun. Any brown dwarfs and their satellites attached to such a former proto-star will be released to form their own planetary nebula while finding their own way in space.

This, then, is the possible mechanism for how a separate brown dwarf planetary system is formed within an overall axially aligned Herbig-Haro object. According to this scenario, as the main proto-star goes main-sequence, we are left with a separate planetary nebula formed far down along the former proto-star's polar axis; a planetary nebula whose planets will likely still be aligned according to the polar configuration required by the Earth/Saturn relationship of mythology.

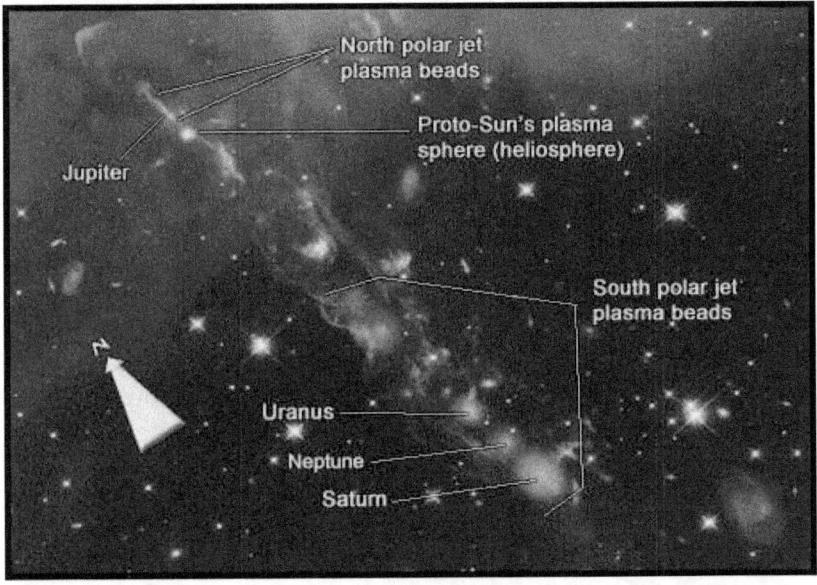

Speculative mock-up composite image showing a perspective view from south south-west of the Sun in its proto-star stage within a Herbig-Haro object. The proto-brown dwarf Saturn and its proto-planets are encased in the plasma bead at the extreme end of the Sun's southern polar jet (Birkeland current), while proto-Jupiter is portrayed as being part of the northern polar jet's string of plasma beads. The distance from the Sun to Saturn would be approximately 1,800 AU, or 15 solar systems away. Image composited from NASA images of HH-34 and HH-110 Herbig-Haro objects. Additional detail added by the authors.

Saturn as the Master of a Free-Floating Planetary Nebula

In the case of Saturn, we would suggest that three plasma beads attached by a Birkeland current to their parent proto-star (the Sun) became separated and formed a string of free-floating proto-brown dwarfs, Saturn being the most southern and the largest of these beads. The two remaining beads north of Saturn reduced to became the planets Neptune and Uranus while continuing to remain in axial alignment with Saturn, becoming trapped in the northern flow of Saturn's dominant Birkeland current. The planets Mars and Earth, we suggest, were ejections from Saturn's core and they remained trapped in z-pinches along the tornado-like flow of the Saturn's powerful southern polar Birkeland current. This Saturnian system, a planetary nebula in its own right, then broke free of the confines of the overall Herbig-Haro object it had been a part of and proceeded to widely spiral northwards through interstellar space. Eventually it would 'catch-up' to its parent proto-star, the Sun, and experience multiple interactions with the Sun's relatively positively charged electrical sphere. Such brushes of contact between the now two different systems would have been played out over multiple times before the Saturnian system was eventually fully captured by the Sun, which had previously captured those northern proto-planets formed in its northern polar jet.

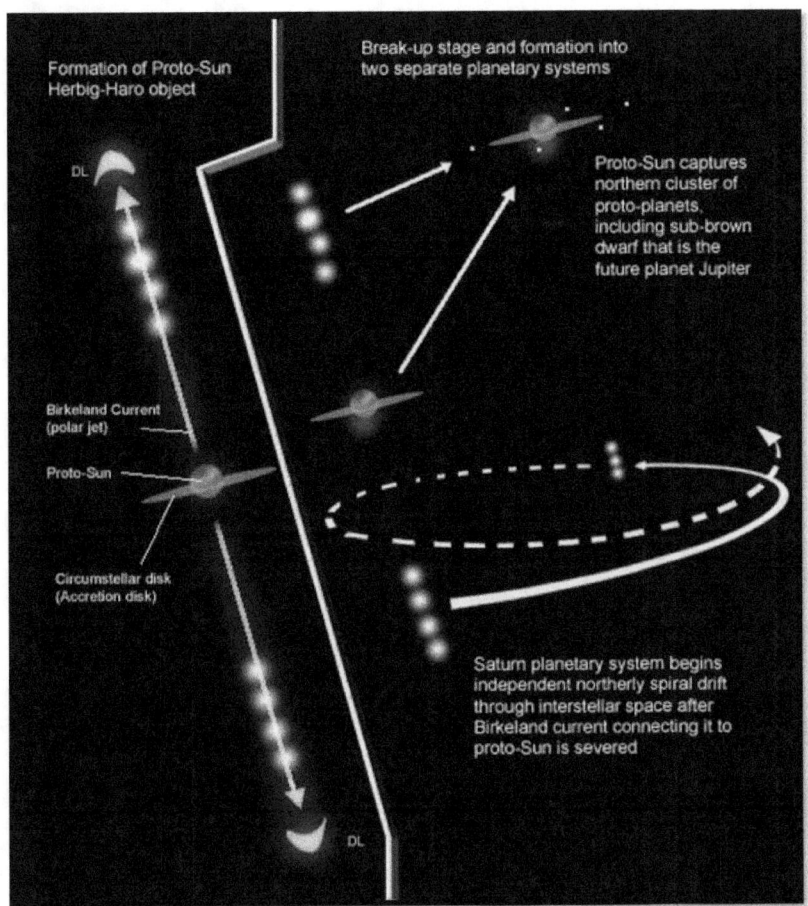

Two stages in the formation and break-up of a Herbig-Haro object with the Sun as its central proto-star. The Sun will quickly catch up with its northern proto-planets to form a classic solar system, while the southern proto-planets will remain in axial alignment with the proto-brown dwarf Saturn. This axial alignment for the Saturnian system of planets is maintained until it caught up to make eventual contact with northern solar system forming around the Sun.

Many of the space objects identified by astronomers as 'planetary' nebulas are, in fact, centers for young proto-stars. The classic signs of electrical Birkeland current activity can be seen in many of these planetary nebulas; for example, signature effects like z-pinches and Double Layers (DLs), which belie the existence of cocooning plasma sheaths, can be seen interacting with the greater interstellar electric field. The possibility that active brown dwarf stars can themselves form these spectacular displays of plasma activity also holds true, especially during periods of enhanced electrical input or disruption. This is what we believe happened to Saturn and its satellites as they found themselves severed from their main proto-star.

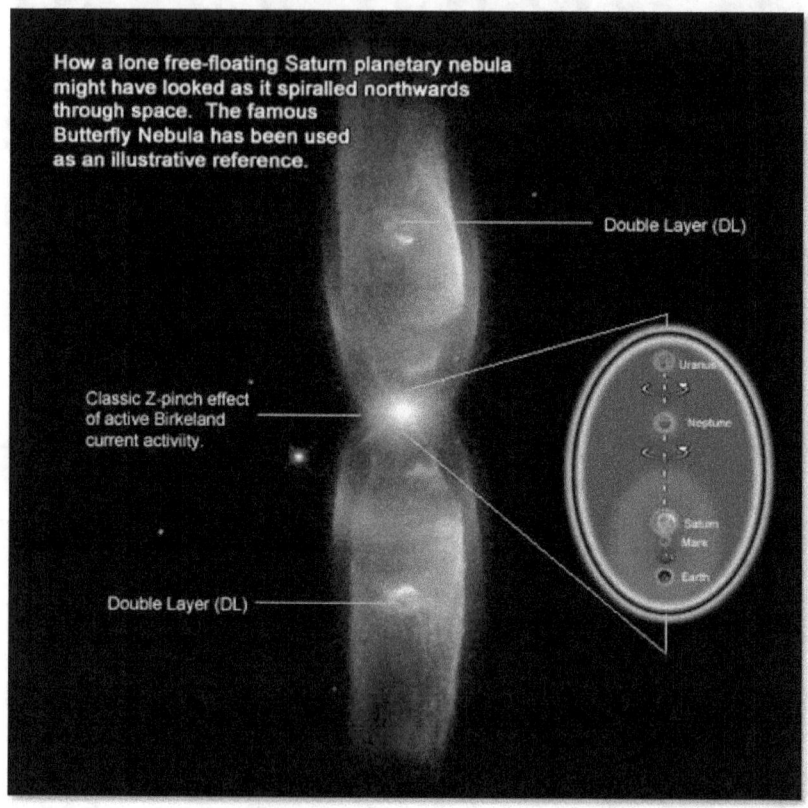

How a lone free-floating Saturn planetary nebula might have looked as it spiralled northwards through space. The famous Butterfly Nebula has been used as an illustrative reference.

Double Layer (DL)

Classic Z-pinch effect of active Birkeland current activity.

Double Layer (DL)

Uranus
Neptune
Saturn
Mars
Earth

After separation from the main proto-star (the Sun's) Birkeland current, the proto-Saturnian system of polar-configured planets would have been encased in their own nebular plasma environment powered by Saturn's own Birkeland current. It is in this environment that mankind experienced the purple dawn era of myth.

Each planetary nebular takes on its own electro-visual characteristics, but the famous Butterfly Nebula (seen above) serves as an illustrative reference for how the Saturnian system may have looked to any hypothetical imaging telescopes seeing it from the depths of space. NOTE: The image of the Butterfly Nebula is a false-color image that can't be detected by the human eye without

After separation, the proto-Saturnian system of planets would have then found itself spiraling independently towards its original proto-star that had formed the Herbig-Haro object that had given it birth. By this time, that very same proto-star had now developed into a full-blown main-sequence star with its own set of orbiting planets, and almost definitely one companion sub-brown dwarf star. That former proto-star is what today we call the Sun. The Sun's companion sub-brown dwarf we are talking about was destined to become the future planet Jupiter, but that part of the story is for the second part of this book.

The primordial and dark 'Purple Dawn' era under proto-Saturn during its sub-brown dwarf star stage. A view looking north from the Siberian arctic coast.

Summary and Takeaways from this chapter

Before light came into the world, humankind's earliest memories are of a primordial and permanent darkness permeated by a dull purple-hued twilight glow. In this 'dreamtime' or 'purple dawn' humanity is said to have had no way of calculating the passage of time since neither the Sun, the moon or the stars could be seen. Instead, the weak and dull glow of a future creative force hovered over a celestial ocean that milled chaotically above humanity for a period that stretched back into an unknowable antiquity.

- The sole energy source for the dull glow that permeated the Purple Dawn era of humanity is recorded as having occupied the northern celestial realms where it maintained a semi-stationary position where the Pole Star is seen today.
- We argue that descriptions of this dark primordial twilight actually describe the conditions of life under a sub-brown dwarf star and that this sub-stellar object would later come to be known as the planet/god Saturn, a celestial body that ancient mythology identifies as Earth's first and best sun.
- Earth was originally a satellite of Saturn during the latter's phase as a sub-brown dwarf star with Earth being cocooned in Saturn's opaque plasma sheath where it received a uniform radiated energy from Saturn in the blue/red spectrum that was reflected back off Saturn's plasma sheath, a type of electrical cocoon.
- The Earth, during its time under Saturn, would have experienced a season-less climate where flora grew in predominantly reddish hues and animal life was primarily adapted to a nocturnal existence (with the notable exception of humans). All light from stars outside of Saturn's plasma sheath would have been blocked from reaching the Earth, thus depriving any humans at that time from being able to calculate the passage of time.
- The predominant force governing the universe (and therefore Earth's primordial relationship to Saturn) is not gravity, but electricity. The transmission of electrical currents via cosmic plasma-based Birkeland currents, where the phenomenon of the electrically-induced and powerful z-pinch can form stars and planets, supersedes gravity as the main force shaping our cosmos.
- Earth was formed and trapped in the z-pinch of proto-Saturn's southern polar Birkeland current where it maintained axial alignment with its original host star in what is called a 'polar configuration'. This axial

82

arrangement persisted and extended back in time for the duration of the period we call the Purple Dawn and only came to an end after Saturn's capture by the Sun. An axial alignment of this type perfectly explains Saturn's stationary position at Earth's celestial north as recorded in world mythology.

- Axial aligned objects are now commonly seen throughout the galaxy where they are often referred to as Herbig-Haro objects. These Herbig-Haro objects are associated with proto-stars that display vast polar jets (Birkeland currents) populated with plasma beads along their length (z-pinch regions where planets can form). Herbig-Haro objects offer a new axial-aligned model for the birthing of brown dwarfs and planets as polar ejected bodies originating from the cores of young proto-stars.

- We hypothesize that our current Sun started out life as a proto-star at the centre of a Herbig-Haro object where its southern polar Birkeland current formed the proto-brown dwarf Saturn and the proto-planets Uranus, Neptune, Mars and Earth, while its northern polar Birkeland current formed Jupiter, Mercury and Pluto.

- We hypothesize that the proto-Sun's axially aligned Herbig-Haro configuration broke up once the Sun went main sequence and that the Sun's northerly drift first allowed it to capture its northern proto-planets of Jupiter (and its moons), Mercury and Pluto before being eventually caught-up by the southern proto-planets dominated by Saturn.

- We hypothesize that in the interim between the break-up of the Sun's proto-Herbig-Haro configuration to its capture of Saturn, the Saturnian collection of proto-planets (including Earth) maintained their axial alignment while drifting in a wide northerly spiral towards the Sun. During this time Saturn and its axial-aligned satellites formed a planetary nebular within an opaque plasma cocoon capable of blocking out all incoming galactic light — i.e. the age known as the Purple Dawn of Mankind.

The Arrival of Man on Earth

The Cold Reality Facing Primordial Mankind

There is no doubting the antiquity of life on Earth as a whole. We can trace the lineages of a huge range of species back through the fossil record to periods vastly predating the arrival of human beings. But at some time in the relatively recent past, the species called Homo sapiens appeared on this planet — and the world has never quite been the same. . .

We jest, but while evolutionists persist in the greater joke of tracing human lineage down through a disconnected and convoluted chain of hominid species to a common simian ancestor, what can be established is that full Homo sapiens (modern man) made his earliest mark with the arrival of a people once referred to as Cro-Magnons.[51]

The world initially inhabited by this Cro-Magnon people was the world of the purple dawn era as discussed in the previous chapter; an age of darkness spoken of in the myths and oral traditions of the most ancient peoples on Earth. As already discussed, it was a dark world imbued with only one season, one climate and absolutely no reference for Cro-Magnons to mark the passage of time. It was also a time when a large chunk of the Cro-Magnons' world was covered by an eternal shadow, a great thick dusty aurorae covering that produced a ribbon-like sheet of ice circumnavigating the Earth's Arctic Circle. It was a time science calls the Pleistocene Ice Age.

---

[51] See Chapter entitled "The Origin of Modern Man: What Were the Requirements?" for the start of a fuller discussion on this topic.

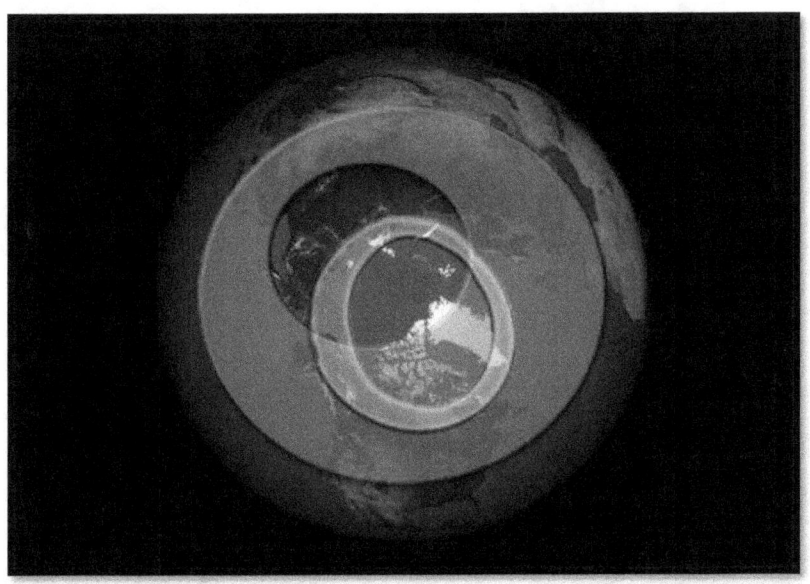

The extent of the Pleistocene ice age is represented by the larger, grey overlay. It clearly shows the Arctic region ice-free, as would have been the case had a sub-brown dwarf star existed for a period of time above Earth's North Pole. The smaller, green colored overlay indicates the general coverage of today's aurora bands. Under Saturn, Earth's aurora band would have been far larger and much denser, a feature that allowed it to cast a cooling shadow over the Earth's surface below them. Such an aurora band probably corresponded to the area covered by the Pleistocene ice sheets.

Land of Eternal Shadow

Dwardu Cardona has postulated that, as Saturn approached the Sun with Earth in its tow, the Saturnian system of planets would have 'electrically sensed' the more positive charge of the Sun's heliosphere as the electrically alive plasma sheaths of both systems drifted towards each other. The net result would be a surge of enhanced electrical activity throughout the Saturnian system, its most noticeable effect on Earth being the emergence of great auroras appearing around the Arctic and Antarctic polar regions.

Auroras are a recognized form of plasma and plasma is known for its ability to attract large concentrations of dust and other particles into cloud-like formations. Due to the close proximity of Earth to Saturn, this aurora would have been highly enhanced compared to the auroras we see today, enhanced enough to have attracted into a vast ribbon of cloud much of the dust in the Earth's atmosphere at that time. This auroral ribbon of semi-permanent cloud would have corresponded very well to the known areas of glaciations that occurred during the Pleistocene ice age.

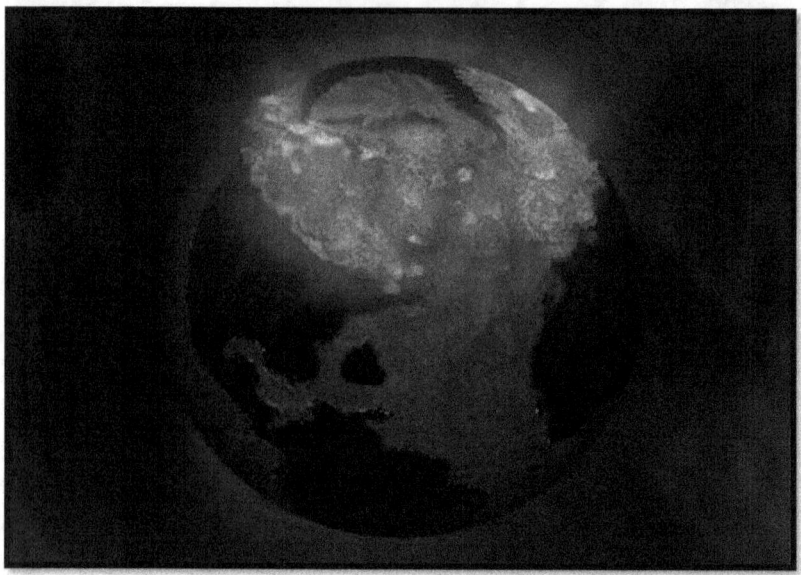

The cooling effect of a dust heavy aurora ring caused by enhanced electrical activity between Saturn and Earth as Saturn starts to 'electrically sense' the more powerful Sun before its capture. This produced the glacial ice sheets that conform to the known glaciated areas of the Pleistocene Ice Age (see previous graphic). Also corresponding to Pleistocene ice coverage, the northern polar region remains ice-free largely due to the concentration of radiated energy arriving from Saturn sitting above the North Pole. The Earth's vegetation takes on a reddish hue at this time giving Earth its purple-hued look for anyone seeing it from orbit.

A common misnomer surrounding Earth's ice ages is that they initially spread from the poles outwards to the lower latitudes as the global temperature cooled. In fact, one of the great mysteries to modern science is that the greater amount of land found within the Arctic Circle remained ice-free during the Pleistocene epoch, a finding that is completely at odds with the accepted model for how ice ages start.

"The islands of the Arctic Archipelago were never glaciated. Neither was the interior of Alaska," wrote R. F. Griggs in an article quoted by Cardona in his book Flare Star,[52] while also quoting Immanuel Velikovsky's reference to J.B. Dana who wrote: "It is a remarkable fact that no ice mass covered the lowlands of northern Siberia any more than those of Alaska."

Cardona also quotes the catastrophists DS Allen and JB Delair (see their work *Cataclysm,* p. 39) in making a case for an ice-free Arctic:

> "Today, the world's coldest known land region is north-eastern Siberia. There, if anywhere, we might expect huge ice-sheets to have developed if the Ice Age theory possessed validity. Yet comparatively very few areas of Siberia exhibit signs of significant glaciation, either past or present.
> . .
>
> "Interestingly, as in neighbouring Alaska to the East, thin rock pinnacles still stand unglaciated at several Siberian localities which thick ice, had it once existed, would unquestionably have ground down and demolished."[53]

And even the venerable *National Geographic* has chimed in on this theme:

---

[52] Dwardu Cardona, "Flare Star," Traffid Publishing, Victoria, B.C., Canada, 2007, page 81; quoting R. F. Griggs, "Indications as to Climate Change From the Timberline of Mount Washington," Science, Vol. 95, No. 2473 (1942), p. 519

[53] Ibid, pp. 81 – 82.

"Ice held most of the northern latitudes in its grip 18,000 years ago – with important exceptions. In the last ice age glaciers never completely covered eastern Siberia, Alaska, and the Yukon." ("Plant that Beat the Ice Age," *National Geographic*, March 2001)

What this all points to is a source of heat radiating from above the north pole at a time when Cro-Magnon man walked the earth during the Pleistocene Ice Age, a source of heat that is well explained by the presence of a radiating, yet dull brown dwarf star that would eventually become the planet Saturn.

The ribbon of glaciated ice sheets that ringed the Arctic Circle would have formed a natural barrier between flora and fauna both north and south of this ice ring, with the majority of Cro-Magnon people settling south of the ice sheets. Yet, it has long been known that lush vegetation once existed within the Arctic Circle, a fact referred to by the writer William Warren in his search for evidence that man's original mythical Paradise had been near the North Pole:

"The Arctic regions, probably up to the North Pole, were not only free from ice, but were covered with a rich and luxuriant vegetation."[54]

Evidence of early humans has also been found there.[55] It seems Cro-Magnon man may have braved crossing the broad glaciated barrier and settled in the areas close to the Arctic coast, and possibly even on some of the Arctic islands. Here, tribes of Cro-Magnons would find themselves closer to the source of life that shed its pale

---

[54] W. F. Warren, "Paradise Found," (Boston, 1885), page 85, as quoting Croll, *Climate and Change* (American edition, 1875), page 7. Both references taken from Dwardu Cardona, "God Star," page 361

[55] John Noble Wilford, "New Evidence of Early Humans Unearthed in Russia's North," September 6, 2001, *The New York Times*, Science section, see; http://www.nytimes.com/2001/09/06/science/06TUSK.html Also see evidence for Neanderthal presence near the Arctic Circle: L. Slimak, J. I. Svendsen, J. Mangerud, H. Plisson, H. P. Heggen, A. Brugere, P. Y. Pavlov, "Late Mousterian Persistence near the Arctic Circle," *Science*, 2011; 332 (6031): 841 DOI: 10.1126/science.1203866. See: http://www.sciencedaily.com/releases/2011/05/110513112527.htm

light on this purple dawn of mankind. Here they could have basked under the primordial dark glow of Saturn, even as it drifted unknowingly ever closer to its fateful encounter with the Sun. Here was a virtual paradise for mankind, a land filled with rich game and flora, a world immortalized on the walls of caves by their cousins still living south of the ice sheets. Here then was the mythical world of the Purple Dawn of Creation.

A Fight for Survival

Such a world as described above can be placed as beginning about 30,000 – 40,000 years ago (at least by standard dating schemes) and lasting up until at least 10,000 years ago when Saturn finally and catastrophically found itself captured by the Sun. The beginning of the Pleistocene ice age coincides with the arrival of Cro-Magnon humans and its end marks the beginning of the mythical Golden Age. Cro-Magnons lived and died during the intervening millennia, leaving us a magnificent testimony as to their existence painted on the walls of deep caves and subterranean tunnels.

During this period, however, Cro-Magnons faced a far more immediate threat to their existence than the depressingly dark atmosphere of primordial Earth. It was a menace that was to have long-reaching consequences for the collective psyche of humans as a whole, a menace that was something truly frightening; it would precipitate a struggle for survival that amounted to what was an Upper Paleolithic world war. The menace in question involved a fight to the death with the leading predator of the age, a creature both cruelly intelligent and immensely strong. In the gloomy depths of this age of darkness, Cro-Magnons found themselves facing their greatest enemy, the Neanderthal!

Summary and Takeaways from this chapter

The appearance of Homo sapiens (modern humans) on Earth was heralded by the arrival of disparate groups of people generically referred to as Cro-Magnons. Their world was, at that time, a dark twilight and timeless existence illuminated only by the dull glow of the sub-brown dwarf called Saturn, a time in which nocturnal animals thrived and the planet found itself in the icy grip of the Pleistocene Ice Age.

- Contrary to popular belief, much of the Arctic polar region was ice-free during the Pleistocene, with the ice sheets actually forming a glacial ribbon around the Arctic Circle and not originating from or extending into the Polar Regions.
- The ice-free land within the arctic circle during the Pleistocene was warmed by the radiating glow of Saturn situated at Earth's celestial north and abounded with flora and mega-fauna at this time.
- The ice sheets encountered by the Cro-Magnon peoples existed in an 'eternal shadow' cast by thick dust-laden aurora generated by Saturn's electrical relationship with Earth. These enhanced aurorae occurred as a result of Saturn's *electrical sensing* of the Sun's heliosphere as it spiraled ever closer to its eventual capture. Similar aurorae existed in the southern hemisphere.
- There is now undeniable evidence for human habitation within the Arctic Circle during the height of the Pleistocene Ice Age; a fact only made possible by the warming influence of Saturn's once radiating energy coming from its northern position where the Pole Star is today.
- The period of time in which Cro-Magnon culture dominated stretches (by standard dating schemes) approximately from 40,000 years ago till 10,000 years ago at which time the Pleistocene Ice Age come to an end as Saturn was captured by the Sun and the Golden Age epoch of world mythology dawned.
- During the Pleistocene Ice Age Cro-Magnons faced the greatest threat to their survival in the form of a predatory hominid called the Neanderthal.

The Origin of Modern Man: What Were the Requirements?

Here we shall examine what we view as the two strongest theories that attempt to explain the origin of modern man on this planet. They are Elaine Morgan's "Aquatic Ape" hypothesis and Danny Vendramini's Neanderthal Predation Theory. Both fail to achieve that goal in our view, nonetheless both involve major advances in our understanding of the human condition.

Neanderthal remains began turning up in the late 1820s in Europe and specimens found in the Neander Valley in 1856 were first assigned a separate category. The remains were of immediate interest to Darwin, Lyell, Huxley and other early proponents of the theory of evolution. Nineteenth century authors described the Neanderthal in unflattering terms, for instance Ignatius Donnelly[1]:

> *In another cave, in the Neanderthal, near Hochdale, between Düsseldorf and Elberfeld, a skull was found which is the most ape-like of all known human crania. The male to whom it belonged must have been a barbarian brute of the rudest possible type....   ...the horrible and beast-like proportions of "the Neanderthal skull" speak, with no less certainty, of undeveloped, brutal, savage man, only a little above the gorilla in capacity;--a prowler, a robber, a murderer, a cave-dweller, a cannibal, a Cain.*

It's a safe bet that Donnelly would not have allowed his daughter to marry such a person. Likewise, artistic representations of the Neanderthal in the 1800s and early 1900s were not flattering:

---

[1] Ignatius Donnelly, "Ragnarok, The Age of Fire and Gravel", page 125

Early Neanderthal reconstruction from the start of the 20th century by Marcellin Boule.

Nonetheless, the image of the Neanderthal that the public sees has undergone a substantial change within the past several decades. Common examples include the reconstructions of Jay Matternes, and a number of others:

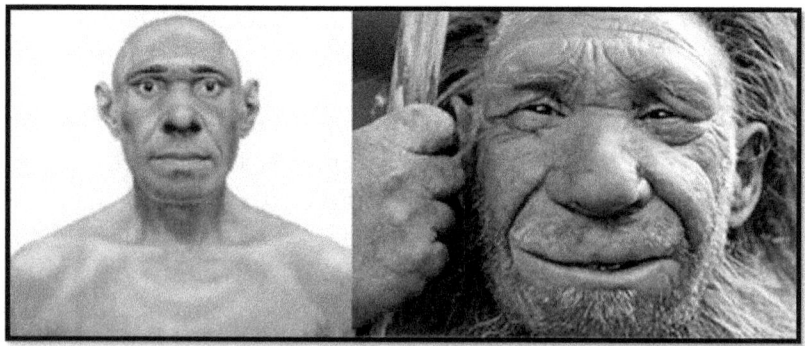

Jay Matternes' Neanderthal reconstruction (left), from the October 81 issue of Science. (Right) Neanderthal reconstruction from an article in The Observer, 16 May 2009.

The two above gentlemen, while certainly different enough, are not exactly frightening. The Neanderthal, thus in recent times, has become a kind of a poster child for what one might call a sort of a multi-cultural approach to evolutionism. Nonetheless, there are still a number of things distinctly wrong with this picture.

One such problem is that if you try to draw such a human-like Neanderthal with the eyes and nose as large as the bones indicate they would have to be, what you end up with is outlandish:

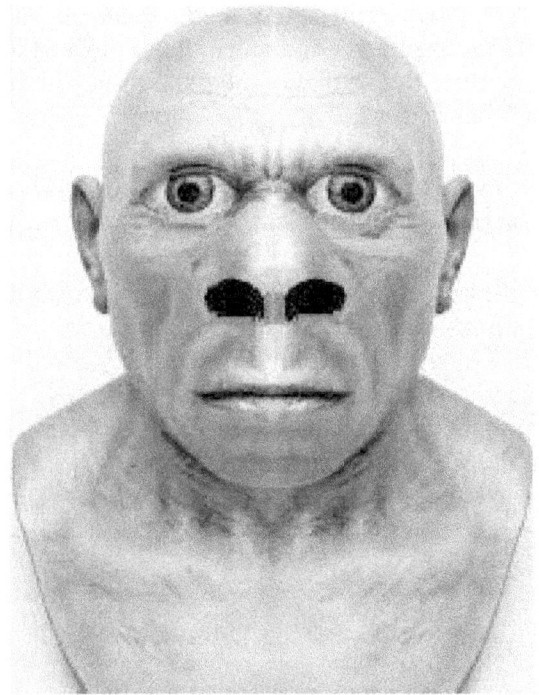

Image courtesy of Rob Gargett, The Subversive Archaeologist Blogspot.

Rob Gargett, the creator of the above reconstruction, notes the following: "Neanderthals: you could possibly have babies with them, but you might want to get out of the way when they sneeze."[2] He then goes on to comment on comparisons between a Neanderthal

---

[2] See:
http://thesubversivearchaeologist.blogspot.com/2011_11_20_archive.html

skull and a human skull and a lion skull (see following image).

"So, I thought I'd do a wee comparison between a modern day "top" carnivore and our cousin's, the Neanderthal, face. Do you see what I see in the image below? It looks as if the felid and the Neanderthal face have more in common than either has with the modern human.

"The lion has a keen sense of smell. Which of the bipedal cousins do you think has the better sense of smell? Relative to the rest of the face, the big cat has a nasal aperture that's equivalent in size to that of the Neanderthal. Not so that of the modern-day hominid on the right.

"A cat can spot its prey from 3 km away. Can you? Do you think the Neanderthal could?

"The cat has dagger-like fangs and molar teeth that would put a deli meat-slicer to shame. "Aha!" you might say, "that chap from Forbes quarry couldn't be as effective as the lion — it doesn't have the appropriate dental accoutrements!" Umm. It's possible, isn't it, that all those flint flakes lying about came in handy for more than whittling?"[3]

---

[3] Ibid.

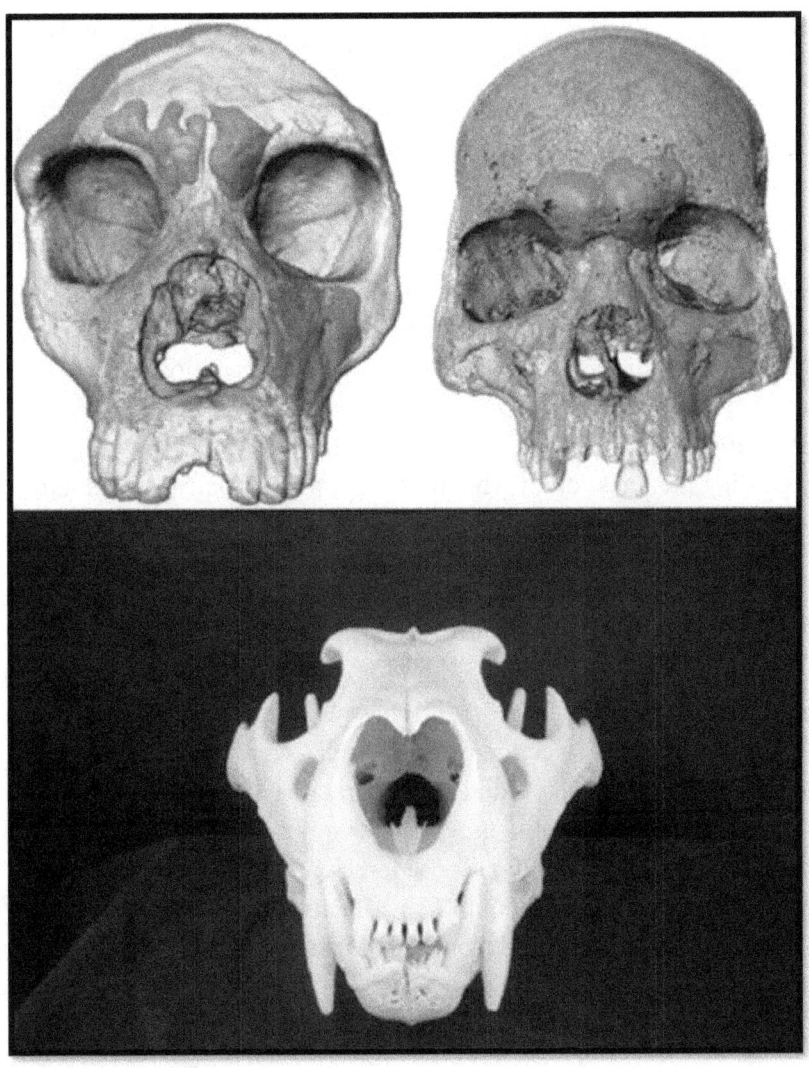

Top: Neanderthal and modern modern human.
Bottom: African lion.
Images courtesy or Rob Gargett,
http://thesubversivearchaeologist.blogspot.com

In other words, much of the Neanderthal's physiology, and
presumably the behavior that one would associate with it, resemble
that of the big cats more than that of humans. Aside from a much
heavier bone structure than ours, Neanderthals had the conical rib
cages seen in primates that allow room for the huge upper body

musculature of the great apes; our own cylindrical rib cages would get in the way of that musculature. It's worth noting too that Neanderthals were hunting large game animals, including mammoths and woolly rhinos, with thrusting spears; not too many humans would sign up for that.

Neanderthals have been viewed as early humans rather than as advanced apes because of the sizes of their skulls and their bipedalism. Nonetheless, studies of Neanderthal DNA in the late 1990s have indicated that Neanderthal DNA was almost exactly halfway between human DNA and that of a chimpanzee[4]. That apparently has sent a lot of people back to the drawing board regarding the question of what these "people" actually looked like.

Vendramini's Neanderthals

Enter at this point one Danny Vendramini, a New Zealand scholar with a multidisciplinary background that includes filmmaking and evolutionary biology. Vendramini has produced a theory featuring a radical new reconstruction of Neanderthals and a thesis claiming that predation by Neanderthals drove gracile (Skhul/Qafzeh) hominids into a very rapid process of evolution such as Stephen Gould and Niles Eldredge proposed, the end result being that those gracile hominids morphed into Cro-Magnon man. Vendramini's study appears to begin with something that somebody should have noticed in the 1800s:

---

[4] E.g. http://expressindia.indianexpress.com/fe/daily/19970712/19355423.html, "He said his team ran four separate tests for authenticity - checking whether other amino acids had survived, making sure the DNA sequences they found did not exist in modern humans, making sure the DNA could be replicated in their own lab and then getting other labs to duplicate their results. Comparisons with the DNA of modern humans and of apes showed the Neanderthal was about halfway between a modern human and a chimpanzee."

This image and Neanderthal reconstructions courtesy
www.themandus.org

That is, aside from the substantially larger brain area, a Neanderthal skull is a near-perfect fit for an ape's profile, while it is a bad fit for one of ours. One assumes also that artists seeking to show the Neanderthal without a "snout" (prognathism), are generally showing Neanderthals looking down at their feet. Vendramini commissioned a forensic artist to create a new Neanderthal reconstruction based on what skulls and bones actually indicate as well as a realistic assessment of the ice age conditions in which Neanderthals lived, the kinds of tools and weapons they used, and the kinds of hunting they engaged in. He did not instruct the artist to create the most frightening monster ever yet seen; apparently he did not have to.

Danny Vendramini's startling take on what a Neanderthal really looked like (Ice-Age fur coat removed for illustration purposes)! Picture courtesy of themandus.org.

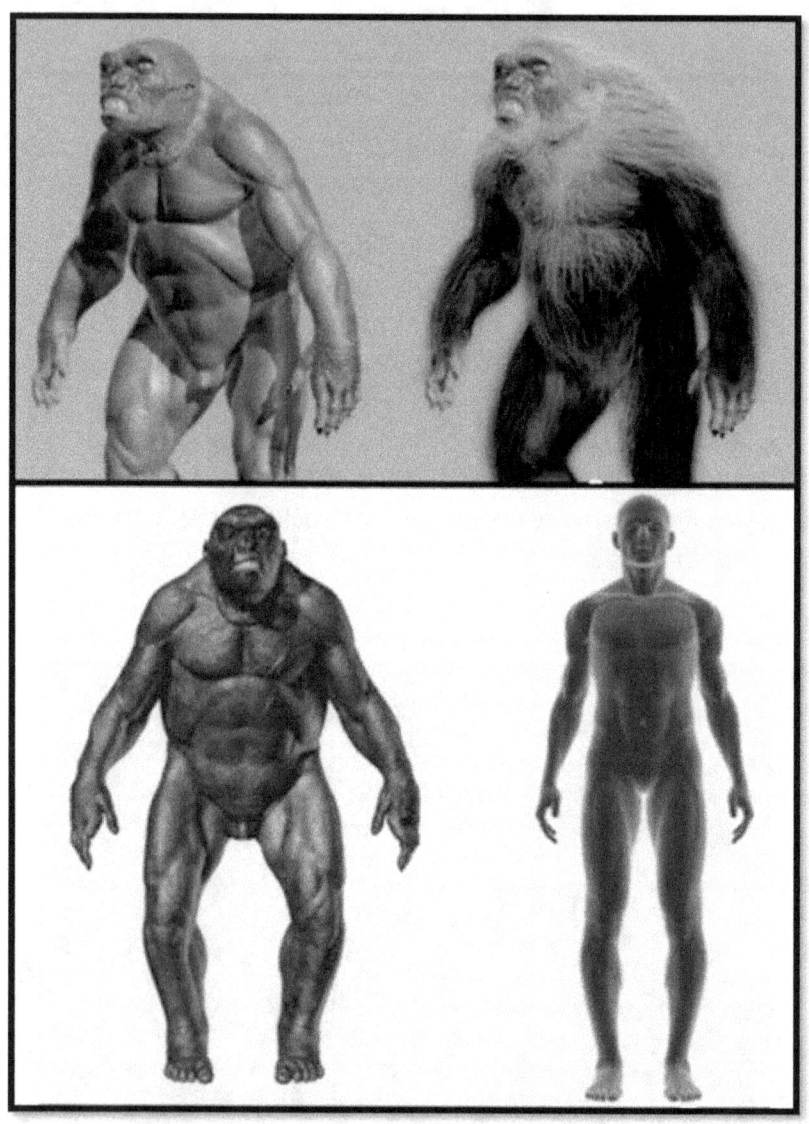

(Above) Danny Vendramini's Neanderthal reconstructions show the Neanderthal with a thick fur coat (top right), while the other images show the creature without it for illustrative purposes. A male

Cro-Magnon needles are not uncommon, while nobody has ever yet found the first Neanderthal needle. A creature with a six inch long ice age fur coat plainly wouldn't have required clothing or needles.

Also, if the comparative human in that last image were about 5 feet and 8 inches tall, then a male Neanderthal could easily be 5 foot eight and weigh 300 pounds or more without any fat on him. That is a creature with physical strength and musculature far beyond anything a human could develop. Lloyd Pye describes the contrast between our bones and those of the Neanderthal as similar to comparing broom sticks to shovel handles[5].

We view Vendramini's reconstructions to be substantially accurate. The two very minor issues we have are the dark gray fur, which we believe should be reddish, and the slit eye pupils, which Vendramini assumes protected the huge eyes in daylight. Due to the placing of Neanderthals in a period of time that we consider consistent with mankind's purple dawn era, we assume that the creature never saw anything that we would call daylight and would subsequently have had no need of the slit eye pupils.

Vendramini's general theory involves several separate claims:

- Neanderthals in Europe morphed into total carnivores and apex predators, while retaining their original primate appearance.

- They developed huge nocturnal eyes and ice-age style fur.

- A band of them found its way into the Levant and began preying on Skhul-Qafzeh hominids, driving the later to the ragged edge of extinction.

- This pressure caused the Skhul-Qafzehs to morph into Cro Magnons as per the Gould/Eldredge version of evolution, in a very short period of time.

---

[5]
http://www.youtube.com/watch?v=pe6DN1OoxjE&feature=player_detailp
age

- Cro Magnons then initiated an upper-paleolithic world war and extirpated all hominids from this planet, root and branch.

Vendramini notes that the Skhul/Qafzeh hominids disappeared from the fossil record, and then the Cro Magnons appeared in the same region; he reasonably enough (to an evolutionist at any rate) assumes that to mean that the population of gracile hominids had been driven down by Neanderthal predation to a very low number, likely under 50 individuals, and then experienced a "speciation event" [evolutionist term for a miracle], which changed them into Cro-Magnons in too short a time period for anybody to measure.

One thing scholars all agree on is that whatever caused Cro Magnon people to appear on this planet when they did was not gradual. Vendramini ("Them and Us") notes[6]:

> "The speed of the Upper Palaeolithic revolution in the Levant was also breathtaking. Anthropologists Ofer Bar-Yosef and Bernard Vandermeersch:
>
> > "Between 40,000 and 45,000 years ago the material culture of western Eurasia changed more than it had during the previous million years. This efflorescence of technological and artistic creativity signifies the emergence of the first culture that observers today would recognise as distinctly human, marked as it was by unceasing invention and variety. During that brief period of 5,000 or so years, the stone tool kit, unchanged in its essential form for ages, suddenly began to differentiate wildly from century to century and from region to region. Why it happened and why it happened when it did constitute two of the greatest outstanding problems in paleoanthropology."

Likewise[7] Dwardu Cardona ("Flare Star"):

---

[6] Danny Vendramini "Them and Us", Kardoorair Press (2009), PDF electronic edition, pages 22 – 23
[7] See: Dwardu Cardona "Flare Star" page 180 - 181

"Where and how the Cro-Magnons first arose remains unknown. Their appearance, however, coincided with the most bitter phase of the ice age. There is, however, no doubt that they were more advanced, more sophisticated, than the Neanderthals with whom they shared the land. Living in larger and more organized groups than had earlier humans, Cro Magnon peoples spread out until they populated most of the world. Their tools, made of bone, stone, and even wood, were carved into harpoons, awls, and fish hooks. They were presumably able hunters although, as with the Neanderthals, they would also have foraged to gather edible plants, roots, and wild vegetables. The only problem here is that, *as far as can be told, the Cro Magnons seem to have arrived on the scene without leaving a single trace of their evolutionary ancestors.* (emphasis ours) 'When the first Cro Magnons arrived in Europe some 40,000 years ago', Ian Tattersall observed, 'they evidently brought with them more or less the entire panoply of behaviors that distinguishes modern humans from every other species that has ever existed.'"

Vendramini's thesis makes sense if you buy into his working assumptions, and if you ignore several problems which we view as insurmountable.

There is the question of people being preyed on by a physically stronger adversary wanting to direct their own evolution towards gracility; in real life, they'd have tried to become heavier and stronger. Vendramini's thesis requires them to become gracile and more adept at throwing things (having the kind of shoulders needed to hurl things) and then invent javelins; in real life, that's needing to get lucky one too many times without having time for it.

There is another problem in that the first evidence of modern humans on this planet includes fine artwork replete with astonishingly good representations of ice age animals on cave walls in multiple colors. Nothing of that sort is associated with any apes or hominids on this planet including the gracile hominids of the Levant. Vendramini claims that the Levantine hominids had spent their final 50,000 years prior to morphing into Cro Magnons, being

101

preyed upon and attempting, consciously or otherwise, to channel their own evolution into a form capable of defeating the Neanderthals. If doing that required mastering fine arts including upper paleolithic versions of the Sistine chapel, then some evidence of it from the 50,000 year run up to this would remain. So far there is no evidence on the planet of hominid artworks.

There is a related problem in that Cro-Magnon art includes self-portraits and those self-portraits are most definitely depicting modern humans and not any sort of a gracile hominid in the process of becoming modern humans (see images below).

Cro-Magnon sculptures (images public domain)

Vendramini describes the change from gracile hominids to Cro-Magnons as being generally driven by a need of the former to differentiate themselves from the latter. The problem is that a number of the things that those gracile hominids would have lost in this process are things that a land prey animal needs to survive in the world; one of the most important being a decent sense of smell, and we view the loss of that is an insurmountable problem for the theory. This presents no problem for Elaine Morgan's aquatic ape thesis of course since an aquatic mammal clearly has little if any need for a sense of smell. On the other hand, for a land prey species to LOSE its sense of smell would be maladaptive and would doom the species. There would be similar questions of such hominids losing their fur while an Ice Age was going on and losing their night vision in an age when night was the only time of day that there was, which we will see in the next chapter.

There is one other insurmountable problem that we discuss in an appendix of this book dealing with evidence for an advanced civilization on Mars in past ages. Suffice it to say that there are a number of face images to be seen on the Martian surface and that those faces are those of modern humans. Vendramini's thesis in combination with the law of averages would require Cro-Magnons to be the first time modern humans had ever appeared in the universe.

We view the evidence that Vendramini presents as indicating that the gracile hominids of the Levant became extinct and that Cro-Magnons shortly afterwards came to the same region. For all of the problems with his general theory, however, we view this theory as the closest which anybody is ever likely to achieve to a workable theory of hominid to human evolution. Vendramini's book contains a great deal of useful information and anybody with more than a passing interest in Neanderthal/hominid topics should have a copy.

Elaine Morgan's Aquatic Ape Theory

The other strong thesis for explaining the origin of modern humans is the aquatic ape theory and the name most commonly associated with that is Elaine Morgan. Morgan is certainly an evolutionist but her theory does not strike us as much of an advertisement for evolution or for evolutionism; she basically lays out reasons for believing that humans are primarily adapted to an aquatic life without offering much of an explanation regarding a process that might have created those adaptations.

Morgan describes a sizeable number of human features that appear to be adaptations for aquatic life and most of which we share with the aquatic mammals. Voluntary control of our breathing is an example. That is basically an adaptation for swimming; we take it for granted but monkeys and apes don't have it, and that is the *only* reason that chimpanzees and gorillas cannot be taught to speak

English; they can be taught to communicate using deaf signs fairly easily.

The most obvious visual difference between us and primates is the fact of our legs being our major limbs; that is basically an adaptation for swimming and wading.

Face to face sex is a behavioral characteristic of aquatic mammals. If any land animals other than humans do this, it's very rare.

Then there is the attribute of human fat, which Morgan describes thusly[8]:

> "The Humans are by far the fattest primates; we have ten times as many fat cells in our bodies as would be expected in an animal of our size.
>
> "There are two kinds of animals which tend to acquire large deposits of fat - hibernating ones and aquatic ones. In hibernating mammals the fat is seasonal; in most aquatic ones, as in humans it is present all the year round. Also, in land mammals, fat tends to be stored internally, especially around the kidneys and intestines; in aquatic mammals and in humans a higher proportion is deposited under the skin.
>
> "It is unlikely that early man would have evolved this feature after moving to the plains and becoming a hunter, because it would have slowed him down. No land-based predator can afford to get fat. Our tendency to put on fat is likelier to be an inheritance from an earlier aquatic phase of our evolution. It is true that some apes, especially in captivity, may put on weight, but we still differ from them in two important ways. One is that they are never born fat. All infant primates except our own are slender; their lives may depend on their ability to cling to their mothers and support their whole weight with their fingers. Our own babies accumulate fat even before birth and continue to grow fatter for several months afterwards. Some of this fat

---

[8] From a web article, http://www.primitivism.com/aquatic-ape.htm. A fuller discussion of human fat characteristics and the adaptation to aquatic life which they represent are found in Elaine Morgan's "The Aquatic Ape Hypothesis", pages 87 - 101

is white fat, and that is extremely rare in new-born mammals. White fat is not much good for supplying instant heat and energy. It is good for insulation in water, and for giving buoyancy.

"The other difference is that in our case the subcutaneous fat is bonded to the skin. When an anatomist skins a cat or rabbit or chimpanzee, any superficial fat deposits remain attached to the underlying tissues. In the case of humans, the fat comes away with the skin, just as it does in aquatic species like dolphins, seals, hippos and manatees."

There was the gigantic problem mentioned above in that virtually all land prey species have senses of smell that are at least adequate in warning them of approaching danger, while the human sense of smell is borderline worthless. That has to be viewed as a fatal flaw not only for Vendramini's theory but for any theory of modern humans evolving on land. However, it is what you would expect given the aquatic thesis for the development of modern man.

Morgan generally describes a large number of such characteristics and, in contrast to the situation with Vendramini's theory, we see only one meaningful problem with Elaine Morgan's (aside from the fact that no evidence of such an aquatic ape has ever turned up): basically, during the time in which man is supposed to have lived on this planet, *there has never been a body of water on this planet* that *would be safe for humans to live in.*

At least as a description of primary human adaptations, the aquatic ape hypothesis is completely believable, other than that it requires a planet other than this one in order to happen. **The planet that the hypothesis requires would be a wet and safe planet, without sea monsters, malarial mosquitoes, crocodiles, hippos, or any of the myriad reasons why humans do not live in water today.** It would also help if the water were predominantly fresh rather than salty.

Humans do not swim as efficiently as fish or other marine mammals and require more energy in order to swim, and that sort of energy generally comes from sugar. You would assume that the original human diet on a safe and watery planet would have been a combination of shellfish, fish, and fruit; and you would assume that the human taste for sweet things arises from the original fruit component of our diet and the primordial need for sugar. Shellfish may have been the more major source of protein; humans can easily deal with shellfish with our hands, while catching fish without the benefit of technology would be more difficult.

The taste for protein and sweet things remains, other food items that have been added to our diet over the ages are less than natural.

The affinity humans have with the ocean has been one of the great factors in sculpting human development since primordial times. Most of our major cities now stand on the shore of some body of water and 'going to the beach' is a worldwide trait amongst humans vacationing. Yet another measure of the extent to which humans still prefer living near water today can be had from the fact that something like 80% of the targets that the US military might ever want to engage were said to have been within the 25 mile range of the guns of the Iowa class battleships, that is, within 25 miles of some shoreline.

Summary and Takeaways from this Chapter

The two versions of a theory of human evolution that come closest to being plausible are those of Danny Vendramini (Neanderthal predation theory) and Elaine Morgan (aquatic ape theory).

Vendramini's reconstructions and descriptions of Neanderthals are perfectly believable. His claim of gracile (Skhul/Qafzeh) hominids being driven by predation pressure from Neanderthals into an accelerated process of evolution into Cro-Magnon people is not believable. There is no evidence of it and too many of the things that would be lost in such a process are things that are altogether necessary for a land prey species to survive. In fact for any hominid to have evolved into Cro Magnon man via *any* version of evolution, that hominid would need to:

- Have lost his fur coat while an Ice Age was going on.
- Have lost almost all of his sense of smell while trying to make it as a land prey species.
- Have lost almost all of his night vision in an age when night was the only time of day to be had.

If that doesn't sound like a formula for success to you, you're not alone. We view Danny Vendramini's theory as the closest anybody is ever likely to come to a workable hominid-to-human evolutionary scheme but, again in our view, it doesn't really work.

If you ignore the question of how apes or hominids might evolve into an aquatic or quasi-aquatic species, then Elaine Morgan's ideas concerning the conditions under which modern humans originally lived seem quite workable; humans actually do share a large number of characteristics with the aquatic mammals. The main problem is that there has never been a body of water on this planet that would be safe for humans to live in.

Morgan's thesis appears to require a planet other than this one; that planet needs to be a wet world and it needs to be generally lacking in aquatic creatures that would be disposed to eat humans.

Eyes for bright worlds and for Dark Worlds

We have seen in a previous chapter that the Sun's solar system originally started off as a double star system consisting of our present sun along with Jupiter and its moons and Mercury and possibly one or more other bodies that are no longer in evidence, and the proto-brown dwarf Saturn along with its planets including Earth and Mars. The moons of Jupiter at that time as well as any other bodies reasonably close to our present sun would have been bright worlds; Earth, Mars, and anything else inside the plasma sheath of Saturn would have been dark worlds. You would expect creatures that were native to one of the dark worlds to be well adapted to a dark world, and you would expect creatures of the bright worlds to be adapted to bright sunlight. We view this as the most major clue as to the origin of modern man.

Consider the eyes of some of the oldest families of creatures on this planet. In fact, the huge eyes may be the single most striking feature in Danny Vendramini's Neanderthal reconstructions:

Most, if not all hominid skulls feature eye sockets notably larger (proportionally) than ours and most, if not all dinosaurs, had the same large-set eyes:

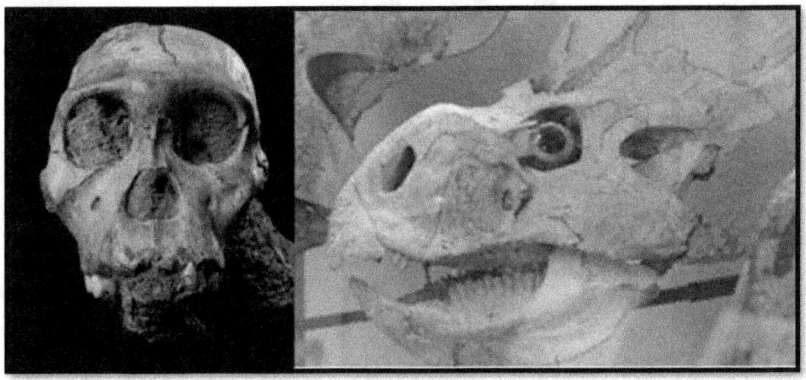

(Above left) Hominid skull showing large eye sockets. (Right) Dinosaur eyes indicate they were nocturnal creatures. Image credit: http://news.discovery.com/animals/nocturnal-dinosaurs-night-fossil-110414.html

The odd thing is that these kinds of eyes appear in ancient herbivores as well as carnivores. In our present world there remain a number of very old kinds of creatures with these kinds of eyes, such as those seen in the collage below:

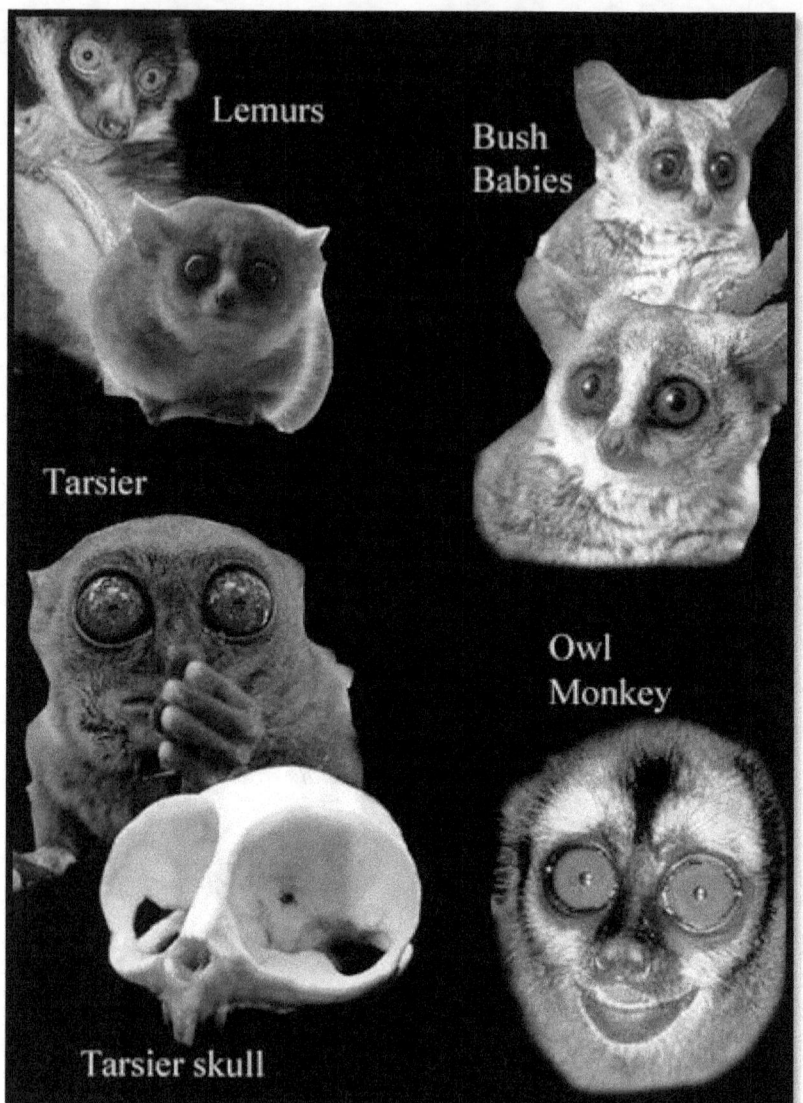

Lemurs

Bush Babies

Tarsier

Owl Monkey

Tarsier skull

Big bright eyes for a dark world. Some of the oldest creatures on Earth display eyes best suited for a dark nocturnal world.

All such creatures that remain are purely nocturnal. But nocturnal traits in our world's present environment do not require those kinds of eyes. Our present world has, in fact, no shortage of creatures that can hunt at night without such plus-size eyesand this includes the big cats.

Today's world is also inhabited by creatures for which sight is a relatively minor sense and for which the main nervous system connection appears to be from the brain to the nose; dogs thus will often flunk a maze test that a rat would easily pass, and then easily find a lost deer by smell. Deer, for their part, see well at night and in daylight see movement well but distinguish forms badly, if at all. It thus turns out that of all the money spent on hunting paraphernalia, the least well spent is on camouflage gear; whitetail deer have been taken by archers wearing white shirts and ties simply seeking to prove to themselves that it could be done. Sight is also a fairly minor sense for hogs so that hunters generally have little difficulty stalking them.

Thus creatures in our present world deal with darkness by various means including eyes better adapted for it than ours as well as the use of other senses. Nonetheless a few leftovers seem to retain the huge eyes of the creatures of past ages, and those kinds of eyes do not appear to be necessary for any sort of darkness our world offers today. In particular, humans and hominids are at opposite extremes of the scale for relative sizes of eyes. **Far from indicating that humans evolved from hominids, this indicates that humans originated in the bright part of our ancient system, while hominids originated in the dark part.** *Humans and hominids do not seem to even originate from the same environment if the sizes of their eyes are anything to go by.* Interestingly, online resources dealing with the sizes of eyes in a variety of species list humans and dolphins as having the smallest eyes relative to the sizes of their bodies:

See: http://www.wonderquest.com/EyeBiggest.htm (Ryosuke Motani):

> Motani comments: "The comparison chart of the figure shows humans and porpoises at the top. Their eye is a barely discernible speck to the right. Squids are next to the bottom of the chart with huge, dinner-plate size eyes."

Our conclusion is that the kinds of eyes seen in hominids were an adaptation to a world that never saw daylight as we experience today. We call these "dark world eyes," while we label our own eyes as "bright world eyes". We assume that the creatures with dark world eyes are native to this planet and originated before the Saturn system was captured by our present sun. We also therefore assume that our own eyes indicate that we cannot be native to this planet, which was dark during primordial times, and must therefore have arisen in a world that was bright in those times.

This realization was hugely counterintuitive at first. Particularly given the Mars/Cydonia findings, which include human face images (see Appendix B), the natural inclination, upon realizing that modern man could not have originated on this planet, was to assume that modern humans may have originated on Mars. Nonetheless, given what we know about the ancient Saturnian system, there is no reason to believe that Mars would have been any brighter than Earth. And what we gather from both the aquatic ape discussion and this observation involving eyes, is that the original home for modern man had to of been a wet world *and* a bright world.

Summary and Takeaways from this Chapter

A number of the oldest groups of creatures on our planet, particularly including dinosaurs and hominids, had/have huge eyes; the eyes are the most striking feature of Vendramini's Neanderthal reconstructions. Those kinds of eyes are what you would expect of creatures living in the darkish world of the Purple Dawn era, when our planet was within the Heliosphere of a sub Brown dwarf star.

Humans and Dolphins on the other hand, have the smallest eyes relative to body size of modern creatures. We are clearly not adapted to the ancient conditions of this planet. Aside from the requirement of the original home world of modern humans being a wet world, as we observe from Elaine Morgan studies, that original home would also have had to be a bright world.

The Upper Paleolithic World War

Predator Versus Warrior

The Neanderthal, in recent decades, has been presented to the public as a slightly different branch of modern man. However, Neanderthal DNA does in fact check out as roughly halfway between ours and that of a chimpanzee and even establishment scientists have dropped the Neanderthal as a plausible ancestor for modern man.

We read, however, that some (not all) modern humans contain Neanderthal genes and that this implies some level of interbreeding between our own ancestors and Neanderthals[1]. Some of these claims come from the Max Planck Institute; nonetheless such claims have to be able to pass simple sniff tests for logic to be believable, and these claims don't.

The claim is that humans of European and Asian descent have 1 – 4 percent Neanderthal genes, and that Africans don't. But there is terribly little genetic diversity amongst modern humans; it is said in fact that there is less genetic diversity in the entire human race than there is in a typical group of 50 African monkeys of the same species. This is attributed to a very recent population bottleneck that the human race went through and that establishment scientists place around 45,000 years ago. At that time, they assume there may have been as few as 100 modern humans alive on the earth.

Any crossbreeding with Neanderthals would have to have been either before or after this bottleneck. If it occurred prior to the bottleneck, Africans would not have been left out. If after the

---

[1] http://en.wikipedia.org/wiki/Neanderthal_genome_project, "In May 2010, the project released a draft of their report on the sequenced Neanderthal genome. Contradicting the results discovered while examining mitochondrial DNA, they demonstrated a range of genetic contribution to non-African modern humans ranging from 1% to 4%. From their Homo sapiens samples in Eurasia (French, Han Chinese & Papuan) the authors state that it is likely that interbreeding occurred in the Levant before Homo sapiens migrated into Europe"

bottleneck, and not involving Africans as per the claim, then the genetic gap between Africans and everybody else would be gigantic rather than minuscule, as it actually is.

Then of course there is the problem of believing that a Neanderthal male could/would rape a woman and, rather than cooking and eating her afterward, somehow or other keep her alive long enough to bear a cross-species child, raise that child to reproductive age, and have him/her breed back into human populations — without anybody catching on.

However, in the real world:

- Neanderthal females would kill that woman the first time her new owner left her alone for ten minutes.

- The woman wouldn't fare any better than the subjects of Soviet attempts to breed humans and apes into super workers in the 1930s.

- Other humans would notice the child was different (the fur coat etc.).

- The humans would kill that child and everybody else like him as part of the same program that killed out the Neanderthal. They would not need DNA tests to determine who to kill for such reasons; it would be exceedingly obvious.

All of the above is before you even get to the untenable position that some scientists have put themselves into regarding the fact that DNA studies have ruled the Neanderthal out as a plausible human ancestor. The claim you read now is that humans and Neanderthals had a "common ancestor," which is normally taken to be Homo Heidelbergensis and assumed to be some half million years in the past[2]. The problem is that "too genetically remote to be ancestral to" is a transitive relationship. In other words and in simplest terms, the claim that the scientists are making is similar to claiming that

---

[2] http://www.guardian.co.uk/science/2012/aug/14/study-doubt-human-neanderthal-interbreeding, ""There was an ancestor of both Neanderthal and modern humans – some archaeologists would call that Homo heidelbergensis – that would have covered Africa and Europe about half a million years ago,""

foxes cannot be descended from wolves because the genetic gap is too wide, and that therefore foxes must be descended directly from fish. Heidelbergensis would probably appear as frightening to a Neanderthal, as a Neanderthal appears to us.

The basic cold hard reality is that we are not related to hominids at all. Genes that we might share with the Neanderthal or any other hominid can as easily be explained as evidence for a designer using a few of the same low-level genetic parts for dissimilar projects. In the same manner, banking software and rocket telemetry software might use a few of the same low-level C language math functions; that clearly does not mean that NASA software is all hacked from JP Morgan's software.

The Neanderthal, whose diet consisted of nearly 100% meat, was the absolute apex predator of ice age Europe. He was an intelligent and tool using creature with massive strength in a compact body who hunted big game animals up to and including mammoths and woolly rhinos with thrusting spears; he appears to have lacked the kind of shoulders needed for throwing javelins or using something like an atlatl. He was a cannibalistic creature[3] (bones are found with unmistakable butchering marks from stone knives) who nonetheless buried his own dead; this indicates that the typical Neanderthal viewed the living world as neatly divided into two categories: his own family group, and meat. One also surmises from this that, aside from making tools and whatever social life existed within the family group, the Neanderthal outlook on life was similar to that of an African lion, and not to that of modern man

Of note is the fact that Neanderthal bones frequently show the kinds of damage that is common amongst rodeo riders. That most likely

---

[3] http://www.sciencemag.org/content/286/5437/18.2.summary, "Neanderthals expertly butchered the game they killed, slicing meat and tendons from bone with stone tools and bashing open long bones to get at the fatty marrow inside. Now, on page 128, a French and American team reports that 100,000-year-old Neanderthals at the French cave of Moula-Guercy performed precisely the same kinds of butchery on some of their own kind. Tantalizing hints of cannibalism have been spotted at other Neanderthal sites for decades, but this is far and away the best documented case, say other researchers."

indicates that they were leaping onto the backs of large prey animals such as mammoths and stabbing them with their spears.

Prior to the arrival of Cro-Magnon man on Earth, the Neanderthal had gone for some very long space of time unchallenged. He was conservative in his habits, his basic tool/weapon kit never changing over his entire existence on Earth. He was not inventive and he didn't get around much. His brain was substantially larger than those of other hominids, a bit larger than ours even, but it hasn't been clear what he used it for; we don't find Neanderthal bridges, aqueducts, or cathedrals, nor even harpoon points or any of the myriad things that Cro-Magnon people made from horn, bone, or stone. New studies are indicating that:

> *"...Results imply that larger areas of the Neanderthal brain, compared to the modern human brain, were given over to vision and movement and this left less room for the higher level thinking required to form large social groups...."* [4]

In other words, the Neanderthal brain was dominated by what you might call the neurological equivalent of the circuitry for a military night-vision scope. This is not surprising, given what we have seen in the previous two chapters.

It is a reasonable conjecture that, prior to the arrival of Cro-Magnons, the only real existential threat to Neanderthals was other Neanderthals. One assumes, therefore, that a lot of whatever creative energy Neanderthals may have had went into dealing, one way or another, with other Neanderthal groups. It could also be, given the tremendous selection pressure there would have been for family group leaders who could keep their fellow hunters alive, that those bigger brains were working overtime designing traps, and that

---

[4] BBC: "Neanderthals' large eyes 'caused their demise'"
http://www.bbc.co.uk/news/science-environment-21759233

the simplest trap may have been more complex than the most complex weapon in those days.

The European Neanderthal appears to have died out in a wave spreading from East to West as he came into contact with Cro-Magnons, with the final Neanderthal stand in Europe occurring in caves on the coast of southern Spain. Recent dating efforts have moved the die-out in Spain back substantially; the picture we used to have of Neanderthals holding out in Spain for three or four thousand years after being extinct in the rest of Europe is being replaced with a picture of a more sudden collapse, ending in Spain[5].

This appears to have been a war of extermination on the part of the Cro-Magnons. There is no reason to believe they would have harbored any hatred for hominids prior to their arrival on Earth, but reasons for such hatred apparently arose in short order, and probably involved Cro-Magnons being eaten. For that matter, lions and hyenas don't always bother to kill prey animals before eating them and there's no particular reason to believe that Neanderthal behavior would have been different. Cro-Magnons would only need to have seen something like that happen to one of their own number once.

On the other hand, if the thought that they needed to exterminate Cro-Magnon man ever occurred to Neanderthals at all, it occurred to them way too late. The greater likelihood is that Neanderthals viewed humans not so much as enemies or adversaries, but as exotic food items. That is to say, that a creature that typically slew bison and mammoths for main course diet staples would not plausibly have viewed humans as a major source of protein; more like something in the way of desert or hors d'oeuvres, and in something like the manner in which we view French pastries. The idea of *exterminating* us would have struck them as abhorrent, much as we

---

[5] http://www.archaeology.org/news/504-130205-spain-neanderthals-extinction, "What our research contributes is that in southern Spain, Neanderthals don't hang on for another 4,000 years compared with the rest of Europe. And the hunch must be that they go extinct in the south of Spain at the same time as everywhere else."

would view the idea of bombing or burning down the French pastry shop.

For a very brief period after the two groups first met, the advantages would have been on the side of the Neanderthals. But this would not have involved anything that you might call a military action; more like groups or hunting parties of several Neanderthals skulking around the perimeters of an early Cro-Magnon settlement until they could find a way to nab one or two humans to carry back to their own settlement for culinary purposes.

Despite the superior physical strength of the Neanderthal however, Cro-Magnons quickly developed several decisive advantages in this conflict. They had javelins, bows, and their signature weapon, the atlatl; they had dogs that would have neutralized any edge the Neanderthal had at night; they had fire and, granted that the Neanderthal also had fire, the Neanderthal with his fur coat had to be a great deal more careful about using fire in the open and could not use it as a weapon. An incident that would cause a minor burn to a human, would most often light a Neanderthal up like a torch and fry him.

But the biggest edge the Cro-Magnons had was probably organization (and the fact that they realized that they were in a war). The large-scale evidence of cannibalism amongst Neanderthals indicates that, as is the case with lions, their social organization was entirely on a family group basis and that trying to organize on any kind of a larger basis than that would have been extremely difficult for them.

All of that says that a Cro-Magnon war party needed be only two or three times the numbers of the largest Neanderthal family they figured they would encounter, i.e. that they never encountered more than one family group worth of Neanderthals at any one time. They could travel with dogs and their projectile weapons and every fifth or sixth person could carry a torch to deal with any Neanderthal that managed to get to close quarters. The whole deal was pretty one-sided and, by the time the Neanderthals realized that they needed anything beyond the organization and skills that had given them

118

dominion over this planet for so many thousands of years, it was way too late.

It is a reasonable conjecture that the Cro-Magnon conquest of Europe took the form of a military invasion rather than merely a sequence of migrations, which would necessarily have included women and children, and that Cro-Magnon's began to settle in Europe after the territory was cleared. One might also surmise that killing Neanderthals was the easy part of the business and that the more difficult task was staying alive in the deep freeze of the European Ice Age. It turns out, in fact, that the extermination of the Neanderthal mainly took place during one of the warmer periods:

> "The data reveals that Team Cro-Magnon began to take over during a not-so-severe climatic era called Greenland Interstadial 8 — an abrupt cold reversal taking place around 40,000 years ago."[6]

Again, the last European stand of the Neanderthal was in southern Spain and you might think that would have been the end of it, but it wasn't. Modern humans spread out from Europe and from the Levant and everywhere they went they exterminated all hominids, and not just the Neanderthal. This gets into areas of investigation at the edge of history and science. Danny Vendramini claims that junk DNA has a way of encoding instinctive behaviors, particularly instinctive recognition of enemies[7]. It is known that turkey chicks will flee from the sight of a hawk and the human abhorrence of spiders and snakes is similarly instinctive.   In fact modelers have flown high-performance radio planes straight out over groups of Canada geese (on the ground) and then turned the planes upwards 5 feet over their heads and the geese don't even blink; that almost

----

[6] Rachel Durfee, "PRESENTING: CRO-MAGNON V. NEANDERTHAL IN THE BATTLE OF EXTINCTION," Popsci.com, 31 December, 2008. See: http://www.popsci.com/scitech/article/2008-12/presenting-cro-magnon-v-neanderthal-battle-extinction

[7] http://thesecondevolution.com/

certainly has something to do with the fact that no goose has ever seen one of his friends killed or eaten by a toy airplane. . .

AXN Floater Jet: Will Canada geese ever come to fear such creatures?

Vendramini claims that human abhorrence of hominids was thus coded into human DNA at an early point so that the war against them continued wherever humans traveled. Interestingly enough, there is no evidence of a human instinct to completely exterminate monkeys or apes. Vendramini refers to this phenomenon as TEEM theory (Trauma Encoded Emotional Memory) and the idea amounts to a claim that Lamarck's vision of evolution actually works for instinctive behaviors

The other question that arises is that of "cryptozoology" and whether or not a certain number of hominids actually remain in the world today. The most densely populated state in the US is New Jersey, which is said to be something like 35% developed. The claim that Lloyd Pye and others make is that something like 65% of the Earth's land surface is seen only from the air and that hominids, including the "Bigfoot" live in such areas. If any such creatures exist, then the fact that no completely verifiable account of them has ever been forthcoming has to mean that previous experiences the creatures have had with humans over the ages have been very, very bad; bad enough that their most basic rule is to avoid contact with us at all costs. Eyewitness accounts of "Bigfoot" describe the creature typically as seven or 8 feet tall, eliminating the Neanderthal from possibility. Amerind oral traditions are not lacking in descriptions of hominids[8].

Hominids came in a number of different flavors. This used to be presented to the public in the form of an evolutionist chain of ascent with presumably remote ancestors at one end of the line leading up to the Neanderthal and then humans at the most recent end. The picture that is emerging now looks more like a case of the various hominid types walking around at roughly the same time. Lloyd Pye claims there are several such creatures still walking around in that 65% of the planet's land surface that is seen only from the air[9].

---

[8] http://www.bigfootencounters.com/biology/fusch.htm

[9]

http://www.youtube.com/watch?v=pe6DN1OoxjE&feature=player_detailpage

Summary and Takeaways from this Chapter

There have recently been claims of genetic evidence indicating crossbreeding between some human groups and Neanderthals. These claims cannot be squared with the dynamics of the recent human population bottleneck and there is no way to believe that early humans would have tolerated hominid crossbreeds at a time during which they were occupied with exterminating hominids, root and branch, from the planet. As James Shreve has noted ("The Neanderthal Peace"), there is no physical evidence of humans and Neanderthals crossbreeding on the planet. In short, we view such claims as false.

The Neanderthal was the apex predator of Ice Age Europe. Compactly built and massively strong, he had dominated the Earth for untold ages before the arrival of Cro-Magnon man, using thrusting spears to kill large prey including mammoths and woolly rhinos. But he lacked the organizational skills and inventiveness of the Cro-Magnons and, viewing them as a food source rather than as an implacable enemy, apparently never realized that he was in a war until it was too late.

Danny Vendramini's claim that instinctive behaviors are heritable ("TEEM theory") appears plausible. If, as Bigfoot aficionados claim, small groups of remnant hominids still live in the 65% of the planet's land surface that is seen only from the air, then their instinct to avoid contact with humans will have arisen, as per Vendramini's claim, from overwhelmingly bad experiences dealing with our ancestors.

Square/Cube Things, and the Ancient Attenuation of Earth's Gravity

Animal and Human Sizes of Past Ages

There is a great deal of interest in questions of giant humans and/or hominids in past ages. Whatever the answer to those questions ultimately turns out to be, there doesn't seem to have ever been an age in which *ONLY* giant humans and/or hominids inhabited our planet. Neanderthals were compactly built, heavy, and immensely strong, but their skeletons don't indicate that they ever grew much over 6 feet tall, if that. Cro-Magnon people appear to have been large, around 6 feet on average, but no more so than groups of larger humans today.

There actually is evidence of substantially larger humans and hominids in past ages but, before getting into that, we should take a look at the question of dinosaurs and their sizes and what made those sizes possible, since that is the simplest case.

The question that you never see in standard textbooks or even in popular literature is this: if those kinds of sizes were such a winning ticket for creatures that supposedly dominated the earth for tens of millions of years, then in the 65 million years that supposedly intervenes between that age and ours, why has nothing else ever re-evolved into such sizes? Or could it be that such sizes are no longer even possible? Or, did you ever wonder why it was always the littlest kid in your class in school who could do the most pushups and pull-ups, or why you never see 200-lb athletes competing in gymnastics?

The answer to all such questions concerning size ratios involves what we call square/cube phenomena, that is, ratios of volume or weight (which is proportional to volume) to some measure of strength or efficiency that is proportional to surface area or to body cross section. Volume, of course, is a cubed figure while cross section and surface area are squared figures. The radiators in cars are basically square/cube phenomena for that matter, since anything only gains or loses heat on its surface; a radiator is a heat transfer device that maximizes surface area for a given volume of coolant.

Weight is proportional to volume, which is a cubed figure (width times breadth times height) while strength is proportional to cross section of bone and muscle, which is a squared figure. Double your physical dimensions, and you have a factor of two that gets figured three times for volume and weight (you'll be eight times heavier), while it only gets figured twice for cross section and strength (and you'll only be four times stronger). You'll have cut your power/weight ratio in half. Clearly you can only halve your power/weight ratio so many times and still stand up and walk; the mathematical limit for that sort of thing in our present world and gravity is about 20,000 lbs., indicating that the largest elephants at something like 15000 lbs. are the largest animals that in actual fact are possible in our present world.

As you get larger, you lower your power to weight ratio no matter what you do. People who work out see this sort of thing in the gym occasionally, and this can be comical. You'll see a reasonably serious weightlifter walk-in with a girl he wants to impress and the first thing that happens (because girls are getting stronger now too) is that the girl walks over to the chinning bar and does 14 or 15 pull ups and the guy figures he has to do 20 or 25; in real life he's going to put himself in the hospital trying to do the 14 or 15 that the girl did and the girl needs some sort of a lecture on how to keep boyfriends alive (i.e. don't allow them to play keep-up with exercises that stress power/weight *ratios*). But back to the case of dinosaurs...

Frederick Malmartel's sketch of a dinosaur addressing the square/cube problem[1]...

It is a fairly easy demonstration (as we'll see shortly) that nothing any larger than the largest elephants could live on land in our world today, and that the largest dinosaurs survived ONLY because the environment on Earth and the structure of the solar system in their age were such that they did not experience gravity as we do today. They would have been crushed by their own weight were they to have experienced our gravity.

And they keep on finding larger and larger dinosaurs. The heavyweight dinosaur crown keeps changing hands with the brontosaurs and brachiosaurs giving way to supersaurs, ultrasaurs, seismosaurs... Christopher McGowan cites a 180 ton weight estimate for the ultrasaur, and describes the volume-based methods of estimating dinosaur weights[2]. He came in for a great deal of criticism for that estimate; more recent estimates are lower but the logic behind McGowan's original estimate has never been refuted to

---

[1] http://frederic.malmartel.free.fr/Fin_des_dinosaures/eedinosaures1.htm
[2] Christopher McGowan, "DINOSAURS, SPITFIRES, & SEA DRAGONS", Harvard, 1991, pp 104-118   McGowan is Curator of Vertebrate Paleontology at the Royal Ontario Museum.

our knowledge. What seems to be the case is that newer sauropod weight estimates are generally low because scientists realize they have a problem.

Argentinosaurus, from Wikipedia article (human in lower left corner of photo):

http://en.wikipedia.org/wiki/File:Argentinosaurus_DSC_2943.jpg

An analysis of dinosaur lifting requirements involves a comparison to human lifting capabilities. One objection that might be raised to this would be to claim that animal muscle tissue was somehow "better" than that of humans. This, however, is known not to be the case:

> "It appears that the maximum force or stress that can be exerted by any muscle is inherent in the structure of the muscle filaments. The maximum force is roughly 4 to 4 kgf/cm2 cross section of muscle (300 - 400 kN/m2). This force is body-size independent and is the same for mouse and elephant muscle. The reason for this uniformity is that the dimensions of the thick and thin muscle filaments, and also the number of cross-bridges between them are the

same. In fact the structure of mouse muscle and elephant muscle is so similar that a microscopist would have difficulty identifying them except for a larger number of mitrochondria in the smaller animal. This uniformity in maximum force holds not only for higher vertebrates, but for many other organisms, including at least some, but not all invertebrates."[3]

Another objection might be that sauropods were aquatic creatures. Nobody believes that anymore; they had no adaptation for aquatic life. Their teeth show wear and tear from eating branches and leaves; you wouldn't get that from eating soft aquatic vegetation. Tracks show them walking on land with no difficulty.

A final objection would be that dinosaurs were somehow more "efficient" than top human athletes, or that they somehow had better "leverage." Superposed images of sauropods and power lifters at roughly equal-weight sizes show the sauropods' legs to be puny compared to the human athletes'. That is not surprising, since a sauropod's body was mostly digestive system (for processing leaves and vegetation), while a human athlete's body is mostly muscle. The better-leverage argument would require the sauropod to be a spectacularly knob-kneed sort of a creature with knees and other joints wider than those of the human athletes, even though the rest of their legs were spindly by contrast. Juxtaposed images of human weightlifters and sauropod dinosaurs at roughly equal sizes do not support this idea.

The normal inverse operator for the square/cube phenomenon is to simply divide by 2/3 power of body weight, and that is indeed the normal (isometric) scaling factor for all weight lifting events.

Suppose for instance that a weightlifting tournament consisted of several different exercises or kinds of lifts (squat, bench-press, deadlift), and that there were several different weight divisions for the athletes of various sizes. For any particular lifting event there would be a best left for each weight division. If you were to divide each of those best lifts by the two thirds power of the corresponding athlete's weight, then the numbers would almost become the same number. One of the numbers would stand out a little bit and that would be the best lift amongst all weight divisions on a scaled basis.

[3] Knut Nielson's, "Scaling, Why is Animal size So Important", Cambridge Univ Press, 1984, page 163

In other words, for human athletes built along similar lines, which includes weightlifters from about 140 to about 220 pounds, the maximal numbers for a particular lift for the champions of the various weight classes all become nearly the same number when divided by the 2/3 power of each athlete's mass. Scaling in this manner eliminates the effect of the athletes' sizes and allows one to determine which athlete, regardless of weight class, has actually achieved the best lift.

The Weightlifter and the Witch

For a quick and dirty look (as opposed to a formal proof...) at how isometric scaling works, let us suppose that John weighs 200 lbs. and can press 300 lbs.

While John is asleep, the wicked witch of the East sneaks into his room and, with her wand, doubles his physical dimensions. As we know, he will be eight times heavier and four times stronger. He now weighs 1600 lbs. and can press 1200 lbs.

Using our calculator, we note that:

$$200^{.67} = 34.8$$
$$1600^{.67} = 140.2$$

and that $\quad 300 / 34.8 = 8.62$
and $\quad\quad\quad 1200 / 140.2 = 8.56$

In other words, dividing by the two thirds power of weight/mass eliminated the difference due to scaling and the square/cube problem and the experiment yielded the same result in both cases within the limits of the two digit accuracy. Basically, dividing by the 2/3 power of mass (or weight since mass is proportional to weight for all cases in this sort of example), simply eliminates the effect of the different sizes.

Again, this sort of isometric scaling is generally used to compare maximal lifting event efforts by athletes of different sizes and works quite well so long as the athletes are built along similar lines. Note however that for any sort of a thought experiment involving scaling the **SAME** athlete to different sizes, this isometric scaling would work perfectly.

Consider the case of Bill Kazmaier, the king of the power lifters in the 1970s and 1980s. Power lifters are amongst the strongest of all athletes; they concentrate on the three most difficult total-body lifts, i.e. bench-press, squat, and dead-lift. They work out many hours a day and, as is fairly common knowledge, use food to flavor their anabolic steroids. No animal the same weight as one of these men could be presumed to be as strong; certainly no herbivore could be presumed to be a strong. Kazmaier was able to do squats and dead lifts with weights between 1000 and 1100 lb. on a bar, assuming he was fully warmed up.

How heavy can an animal still get and survive on land in today's world, then? This amounts to the same thing as asking the question of how heavy Mr. Kazmaier would be at the point at which the square-cube problem made it as difficult for him just to stand up as it is for him to do 1000 lb. squats at his competitive size of 340 lb.? The answer is simply the solution to:

$$1340/340^{.667} = x/x^{.667}$$

Or just under 21,000 lb. In fact, that would be the point at which just standing up would represent the same level of effort is a fully warmed up, one-shot, go for the gold maximal total body lift. In real life, standing and walking have to come more easily than that.

Now, if you were to put a top power lifter like Kazmaier next to a sauropod dinosaur that is equal in size, what you would be looking at would be one animal at the top of the food chain and another at the bottom. The human athlete's body is mostly bone and muscle; the herbivore's body is mostly made up of a digestive system for processing leaves, grass, and other very low value foods. And if Kazmaier couldn't make it past 20,000 pounds in our world's present gravity, the herbivore certainly could not. Again, in all cases, we are comparing the absolute max effort for a top human weight lifter to lift and hold something for two seconds versus the sauropod's requirement to move around and walk all day long with a scaled weight greater than these weights involved in the maximum, one-shot, two-second effort.

Thus, in real life, elephants do not appear to get to that 20,000 lb. point. Christopher McGowan claims that a Toronto Zoo specimen was the largest in North America at 14,300 lb.[4], and Smithsonian

---

[4] Christopher McGowan, "DINOSAURS, SPITFIRES, & SEA DRAGONS", Harvard, 1991 page 97

personnel once provided a weight estimate of about eight tons for the huge bush elephant whose body stands in the Museum of Natural History.

There are other square/cube problems aside from the ability to lift weights. Oxygen consumption (the ability to breathe), the ability to digest food, and wing-loading and hence the ability to fly, are all things that vary with a squared figure and yet must support mass, which varies with a cubed figure, as noted. All of these things impose size limits on animals of various kinds.

How Much Attenuation of Gravity was Needed to Account for Dinosaur Sizes?

Thus we observe and conclude that dinosaurs were only able to exist because gravity was attenuated during their age; the next question is, how much attenuation was required for the largest sauropod dinosaurs to exist?

From physics, weight = mg, or mass times the acceleration of gravity at the Earth's surface. We assume that the Smithsonian elephant at 16,000 lbs. is ballpark for the heaviest creature our world will support today and also that the ultrasaur, for which Christopher McGowen gives a 360,000-lb weight figure[5] (we believe that original volumetric calculation was correct) was the heaviest or near the heaviest of their world. We also assume that the level of effort for the sauropod to stand in his world cannot have been more than it is for the elephant in ours. Using our athletic scaling factor again, we use **gd** for gravity-dinosaur, **ge** for gravity-elephant, **md** for mass-dinosaur, **me** for mass-elephant, and note that:

$$\textbf{me * ge = 16000}$$
$$\textbf{md * ge = 360,000}$$

i.e. we note that, in our gravity, the elephant   weights 16,000 lbs. while the dinosaur would weigh 360,000 lbs., so that:

$$\textbf{md = (360/16) * me}$$

---

[5] Ibid, page 118

We also observe that:

$$(md * gd)/md^{.67} = (me * ge)/me^{.67}$$

(scaled lifts for standing are equal)

$$gd * md^{.34} = ge * me^{.34}$$

$$gd/ge = me^{.34} / md^{.34}$$
$$= me^{.34} / (360/16 * me)^{.34}$$
$$= me^{.34} / (me^{.34} * (360/16)^{.34})$$

Thus $2.8 * gd = ge$, i.e. the ratio of gravity then vs. now is the cube root of the ratio of the weights of the two animals. In other words, *it would take almost a three to one attenuation of the acceleration due to gravity in order for the largest dinosaurs to exist!*

Sauropod Necks and the Problem of Torque

A second category of evidence for the attenuation of gravity in prehistoric times arises from the study of sauropod dinosaurs' necks. Scientists who study sauropod dinosaurs have claimed that they held their heads low, because they could not have gotten blood to their brains had they held them high. McGowan goes into this in detail on this topic[6]. He mentions the fact that a giraffe's blood pressure, at 200 - 300 mm Hg, far higher than that of any other animal, would probably rupture the vascular system of any other animal. This pressure is maintained by thick arterial walls and by a very tight skin, which apparently acts like a jet pilot's pressure suit. A giraffe's head might reach to 20'. How a sauropod might have gotten blood to its brain at 50' or 60' is the real question.

Two articles that mention this problem appeared in the 12/91 issue of Natural History. Harvey B. Lillywhite of Univ. Fla., Gainesville, noted:

> "...in a Barosaurus with its head held high, the heart had to work against a gravitational pressure of about 590 mm of

---

[6] Christopher McGowan, "DINOSAURS, SPITFIRES, & SEA DRAGONS", Harvard, 1991 pp 101 - 120

mercury (Hg). In order for the heart to eject blood into the arteries of the neck, its pressure must exceed that of the blood pushing against the opposite side of the outflow valve. Moreover, some additional pressure would have been needed to overcome the resistance of smaller vessels within the head for blood flow to meet the requirements for brain and facial tissues. Therefore, hearts of Barosaurus must have generated pressures at least six times greater than those of humans and three to four times greater than those of giraffes."[7]

In the same issue of Natural History, Peter Dodson noted:

"Brachiosaurus was built like a giraffe and may have fed like one. But most sauropods were built quite differently. At the base of the neck, a sauropod's vertebral spines unlike those of a giraffe, were weak and low and did not provide leverage for the muscles required to elevate the head in a high position. Furthermore, the blood pressure required to pump blood up to the brain, thirty or more feet in the air, would have placed extraordinary demands on the heart (see opposite page) [Lillywhite's article] and would seemingly have placed the animal at severe risk of a stroke, an aneurysm, or some other circulatory disaster. If sauropods fed with the neck extended just a little above heart level, say from ground level up to fifteen feet, the blood pressure required would have been far more reasonable."[8]

It turns out, however, that a problem every bit as bad or worse than the blood pressure problem would arise, gravity being what it is now, were sauropods to hold their heads out just above horizontally as Dodson suggests. Try holding your arm out horizontally for more than a minute or two, and then imagine your arm being 40' long and 30,000 lb. . .

An ultrasaur or seismosaur with a neck 40' - 60' long and weighing 25,000 – 40,000 lb., would be looking at 400,000 to nearly a million foot pounds of torque were one of them to try to hold his neck out horizontally. That's basically impossible. You don't hang a 30,000 lb. load 40' off into space even if it is made out of wood and structural materials, much less flesh and blood. In fact, if you set

---

[7] Natural History, December 1991: "Sauropods and Gravity", Harvey B. Lillywhite
[8] Ibid: "Lifestyles of the Huge and Famous", Peter Dobson

out to research the question of what, if anything, in our world involves torques on the order of half a million to a million foot pounds, you'll find that no nut or bolt on anything in the world involves more than a few thousand foot-pounds of torque. The only thing in the ballpark would be the combined maximum torque of all the engines of a WW-II battleship or a modern aircraft carrier, that is, sufficient torque to drive a 60,000 ton ship through the water at 30 knots or better. The idea of anything made of flesh and blood holding that much torque is very far removed from reality.

A cursory look at an elephant's skeleton reveals a structural system much like Roman architecture with one and only one purpose in mind, i.e. bearing the elephant's great weight. The legs are columns and the spine is a Roman arch. A sauropod's neck, however, particularly in the case of the recent ultrasaur and seismosaur finds, weighed several times the weight of a large elephant and, if held outwards horizontally, would actually arch downwards
(the **wrong** way). Reconstructions actually depict them like that, no thought whatever given to the consequences, either by the scientists or the artists involved.

And so, sauropods (in our gravity) couldn't hold their heads up, and they couldn't hold them out either. That doesn't leave much.

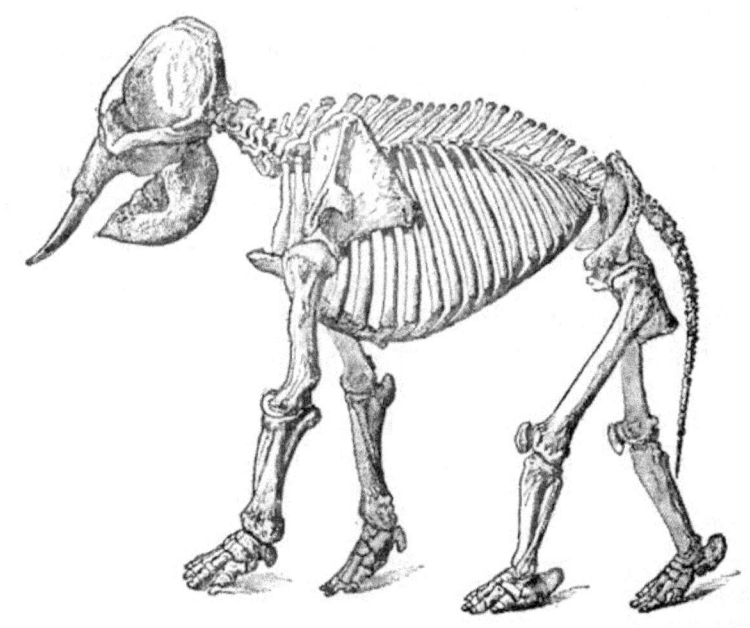

(above) An elephant's body build is termed "graviportal," and consists of the basic elements of Roman architecture, the legs serving as columns, and the spine serving as a Roman arch for supporting weight.

If an elephant's spine were to arch downward as is the case with most quadrupeds, the elephant would collapse. (public domain image[9])

Public Domain Image

Diplodocid (sauropod which held its neck outwards) reconstruction. Notice that the neck arches the wrong way. (public domain image)[10]

Ancient Flying Creatures

A third category of evidence for attenuated gravity in antediluvian times arises from studies of creatures that flew in those times, and of creatures that fly now. In the prehistoric world, thousand-pound flying creatures soared in skies that no longer permit flying creatures above 30 lb. or thereabouts. Modern birds of prey (the Argentinean teratorn) weighing 170 -200 lb. with wingspans of 30' also flew in ages past, while within recorded history central Asians have been trying to breed hunting eagles for size and strength, yet have never gotten them beyond approximately 25 lb. At that point they are able to take off only with the greatest difficulty.

A book of interest here is Adrian Desmond's "The Hot Blooded Dinosaurs." Desmond has a good deal to say about the pteranodon,

---

[9] http://en.wikipedia.org/wiki/Asian_elephant

[10] http://tinyurl.com/cazntcn

the 40 - 50 lb. pterosaur that scientists used to believe was the largest creature that ever flew:

"Pteranodon had lost its teeth, tail and some flight musculature, and its rear legs had become spindly. It was, however, in the actual bones that the greatest reduction of weight was achieved. The wing bones, backbone and hind limbs were tubular, like the supporting struts of an aircraft, which allows for strength yet cuts down on weight. In Pteranodon these bones, although up to an inch in diameter, were no more than cylindrical air spaces bounded by an outer bony casing no thicker than a piece of card. Barnum Brown of the American Museum reported an armbone fragment of an unknown species of pterosaur from the Upper Cretaceous of Texas in which 'the culmination of the pterosaur... the acme of light construction' was achieved. Here, the trend had continued so far that the bone wall of the cylinder was an unbelievable one-fiftieth of an inch thick Inside the tubes bony crosswise struts no thicker than pins helped to strengthen the structure, another innovation in aircraft design anticipated by the Mesozoic pterosaurs.

"The combination of great size and negligible weight must necessarily have resulted in some fragility. It is easy to imagine that the paper-thin tubular bones supporting the gigantic wings would have made landing dangerous. How could the creature have alighted without shattering all of its bones How could it have taken off in the first place It was obviously unable to flap twelve-foot wings strung between straw-thin tubes. Many larger birds have to achieve a certain speed by running and flapping before they can take off and others have to produce a wing beat speed approaching hovering in order to rise. To achieve hovering with a twenty-three foot wingspread, Pteranodon would have required 220 lb. of flight muscles as efficient as those in humming birds. But it had reduced its musculature to about 8 lb., so it is inconceivable that Pteranodon could have taken off actively.

"Pteranodon, then, was not a flapping creature, it had neither the muscles nor the resistance to the resulting stress. Its long, thin albatross-like wings betray it as a glider, the most advanced glider the animal kingdom has produced. With a weight of only 40 lb. the wing loading was only I lb. per square foot. This gave it a slower sinking speed than even a man-made glider, where the wings have to sustain a weight of at least 4 lb. per square foot. The ratio of wing area to total weight in Pteranodon is only surpassed in some of the insects. Pteranodon was constructed as a glider, with the breastbone, shoulder girdle and backbone welded into a box-like rigid fuselage, able to absorb the strain from the giant wings. The low weight combined with an enormous wing span meant that Pteranodon could glide at ultra-low speeds without fear of stalling. Cherrie Bramwell of Reading University has calculated that it could remain aloft at only 15 m.p.h. So takeoff would have been relatively easy. All Pteranodon needed was a breeze of 15 m.p.h. when it would face the wind, stretch its wings and be lifted into the air like a piece of paper. No effort at all would have been required. Again, if it was forced to land on the sea, it had only to extend its wings to catch the wind in order to raise itself gently out of the water. It seems strange that an animal that had gone to such great lengths to reduce its weight to a minimum should have evolved an elongated bony crest on its skull."[11]

Desmond has mentioned some of the problems that even the pteranodon faced at 50 lb. or so; no possibility of flapping the wings for instance. And then:

> **"Calculations bearing on size and power suggested that the maximum weight that a flying vertebrate can attain is about 50 lb.** (emphasis ours): Pteranodon and its slightly larger but lesser known Jordanian ally Titanopteryx were therefore thought to be the largest flying animals."

We are not aware of any demonstration of a fatal flaw in those calculations, other than for assuming that gravity in prehistoric times would have been the same as it is now.

---

[11] Adrian J. Desmond, The Hot-Blooded Dinosaurs: A Revolution in Paleontology, New York, 1976, pages 178 - 183

The giant teratorn finds of Argentina were not known when Desmond's book was written, they came out in the eighties in issues of *Science Magazine* and other publications. The aforementioned teratorn was a 160 - 200 lb eagle with a 27' wingspan, a modern bird whose existence involved flapping wings, and aerial maneuver. And then there is the case of the Texas pterosaurs.

(Image:
http://www.jhu.edu/~gazette/2009/12jan09/12pterosaurs.html)

Robert T. Bakker has this to say about the Texas Pterosaurs:

> "Immediately after their paper came out in Science, Wann Langston and his students were attacked by aeronautical engineers who simply could not believe that the Big Bend dragon had a wingspan of forty feet or more. Such dimensions broke all the rules of flight engineering; a creature that large would have broken its arm bones if it tried to fly... Under this hail of disbelief, Langston and his crew backed off somewhat. Since the complete wing bones hadn't been discovered, it was possible to reconstruct the

Big Bend Pterodactyl [pterosaur] with wings much shorter than fifty feet."[12]

The original reconstruction had put wingspan for the pterosaur at over 60'. Bakker goes on to say that he believes the pterosaurs really were that big and that they simply flew despite our not comprehending how, i.e. that the problem is ours. He does not give a solution as to what we're looking at the wrong way.

In the cases of birds larger than 25 or 30 pounds that survived the change in gravity, that is, in the cases of ostriches and New Zealand moas and the like, the wings became vestigial and the creatures developed lifestyles that did not depend upon flight. Teratorns and the Big Bend pterosaurs, on the other hand, had wings that were clearly not vestigial and could not have lived other than by flying.

One of the problems here is that Einstein's description of gravity as a four dimensional differential geometry kind of thing would not allow anybody to believe that gravity could have undergone any sort of a large change near the surface of the earth in geologically recent times. Nonetheless, as we have just seen, it is an easy demonstration that it must have. What this means is that gravity is not an absolute basic force in nature and must actually be some kind of electrostatic dipole phenomenon as Ralph Sansbury and researchers connected with Thunderbolts.info claim.[13]

Ancient Humans and their Sizes

So much then for dinosaurs and their sizes. Those sizes were only possible because of an attenuation of gravity that prevailed in prehistoric times. What about other kinds of creatures, particularly humans? Is there any reason to believe that at least some humans and/or hominids would have been substantially larger in past ages than might be possible today?

The limit of size for any particular creature depends upon the kinds of stresses that the creature has to be able to deal with and this is because of the square/cube problem that we have mentioned. It is also the case that these limits have been uniformly higher in an age not more than a few tens of thousands of years ago. These included mammoths, which were substantially larger than present elephants, 1500 pound lions and 2500 pound bears in California, an eight-foot

---

[12] Robert T. Baker, "The Dinosaur heresies", Zebra Books, pp 290-291

[13] http://www.holoscience.com/wp/electric-gravity-in-an-electric-universe/

long beaver in New York, and a number of other such mega fauna. But what about humans?

From the time humans first appeared on this planet, there does not appear to have ever been a shortage of humans who were in the ballpark for the kinds of sizes that humans attain to now. Nonetheless, from what we've just seen, it would be surprising to learn that there had never been humans substantially larger than present humans. Most are familiar with the statement from Genesis:

> GEN 6:4 There were giants in the earth in those days; and also after that. . .

But there does not appear to be any physical evidence of giant humans from past ages... Does that mean that there is no such evidence, or simply that all or nearly all such evidence has been buried or covered up? Conspiracy theories should normally be viewed as a last resort, for situations in which all other remedies have failed; however we view this issue as a legitimate case involving conspiracy to cover up.

It was common knowledge in the United States in the 1700s and 1800s that there were burial mounds at numerous sites containing the bones of giant humans, and even Abraham Lincoln mentioned this once in a speech given at Niagara Falls:

> "But still there is more. It calls up the indefinite past. When Columbus first sought this continent---when Christ suffered on the cross---when Moses led Israel through the Red-Sea---nay, even, when Adam first came from the hand of his Maker---then as now, Niagara was roaring here. The eyes of **that species of extinct giants, whose bones fill the mounds of America**, (emphasis ours) have gazed on Niagara, as ours do now. Co[n]temporary with the whole race of men, and older than the first man, Niagara is strong, and fresh to-day as ten thousand years ago. The Mammoth and Mastadon---now so long dead, that fragments of their monstrous bones, alone testify, that they ever lived, have gazed on Niagara. In that long---

long time, never still for a single moment. Never dried, never froze, never slept, never rested."[14]

Prior to the Internet age, the educated layman had little opportunity to read or examine any of this type of material. Nonetheless, the age when committed ideologues could keep lids down on this sort of thing is over. We recommend the following two starting points for interested readers:

Chapman Research Report:

http://www.chapmanresearch.org/PDF/There%20Were%20Giants%20on%20the%20Earth.pdf

Robert Vannrox post on FreeRepublic Forum from email from Vine Deloria:

http://www.freerepublic.com/focus/f-news/720497/posts

Articles that attempt to describe the role of the Smithsonian Institute in these areas include:

http://jmilor.startlogic.com/articles/The%20Giant%20Conspiracy.html

http://www.burlingtonnews.net/smithsonian.html

The Chapman Research Project and other similar resources that we have examined document numerous giant human finds, and have several recurring themes or elements in common:

- The giant human remains being described involved humans from 6 1/2 to around 13 feet tall; there no reports of 50 or 100 foot tall humans.
- A number of the reports note that the remains of giant humans appeared to be the remains of healthy people and not people suffering from glandular diseases as is the case with any human over 8 feet tall in the modern age.
- The most common sizes are in the 7 to 8 foot tall range.
- By far the greatest number of such reports originate from North America.

---

[14] "Collected Works of Abraham Lincoln. Volume 2," pages 10 – 11. See http://quod.lib.umich.edu/l/lincoln/lincoln2/1:6?rgn=div1;view=fulltext

- The greatest number of such reports appear to originate from the 1880s. That coincides with the great age of building in the United States, the perfection of the modern steel industry, the availability of dynamite and other advances in engineering, all of which tended to make the occasional archaeological find more likely.
- A number of the reports in the Chapman Project include claims of double rows of teeth. This would be an adaptation to a substantially longer lifespan than currently enjoyed by modern humans.
- Many of the reports indicate that bones of a very great age, within a few days of being exposed to air, crumbled into dust.
- Most reports describe finding the bones of giant humans; a few mention archaic characteristics, possibly indicating hominids rather than humans.

Later in this work, we discuss the idea that there were at least two separate saltations of modern humans on this planet. There is enough evidence for diverse groups of humans having concurrently been present on Earth in past ages that we cannot rule out the possibility that there may have been more than two saltations.

Antediluvian Lifespans

There is one other thing to note about attenuated gravity and the effect that it would have had on humans. Within recorded history, the ratio of maturity to lifespan for humans has been something like 1 to 4, that is, men would marry and begin families typically at an age of physical maturity, roughly 20, and live to a maximum of something like 80 years. Granted it is not possible to know exactly what the word "year" would have meant prior to the Biblical account of the flood; nonetheless the ratio was different. Genesis describes people living to 60 or 70 of whatever they called years, marrying and having first children, and then living to 800 and 900 years more or less. That amounts to a ratio closer to 1 to 13.

Gravity is the major cause of physical stress in the world and the reason that human bone structures begin to collapse and people begin to die of arthritis-related problems if they have not already died from anything else prior to that. For that matter, most neurosurgeons and orthopedists would agree that there is no such thing as a totally healthy human back past forty years of age. This would have to make anybody wonder whether humans could be said to be adapted to life on this particular planet as it exists at present, or whether in fact modern humans would have arisen at all in our present world.

In a world in which gravity was no more than a third of its present value (the largest sauropod dinosaurs would need at least that much attenuation simply to stand), you would expect humans to lead longer and healthier lives. And, under such circumstances, one has to assume that an advanced civilization could arise much more rapidly than has been the case within our own recorded history. The best minds in such a society would have a great deal more time in which to work.

That doesn't mean that ALL humans living in prehistoric societies would have been spacefaring or otherwise highly advanced; there would have been groups of humans at various levels of technological development then, even as there are now. Nonetheless it goes a certain ways towards understanding how spacefaring civilizations could have arisen in prehistoric times without taking hundreds of thousands or millions of years to do so.

Summary and Takeaways from this Chapter

Weight is proportional to volume, a cubed figure, while strength is proportional to cross section of bones and muscle, a squared figure. Double your physical dimensions, and the factor of two will be figured three times for volume and weight (you'll be eight times heavier), while cross section and strength, which vary like πr^2, will only square the factor of two (you'll be four times stronger). You'll have cut your power to weight ratio in half. Clearly, any creature can only halve its power to weight ratio so many times and still stand and walk. This so-called square/cube problem imposes a size limit for living creatures for the planet; at present, that limit is around 20,000 pounds.

Sauropod dinosaurs would be crushed by their own weight in our present world; a nearly 3-1 attenuation of the present acceleration of gravity would be minimally required for the largest sauropods to stand at all.

Evidence of giant humans from past ages was apparently common and commonly known in the 1700s and 1800s with no less than Abraham Lincoln referring to them in a speech given at Niagara Falls. References that you see an old newspaper and Journal articles most often refer to human remains indicating heights of seven, eight, or nine feet or thereabouts. There does not appear to have ever been a shortage of humans of more normal sizes on the planet and there are no references to humans 30 feet tall or 100 feet tall. It is not unreasonable to suspect foul play (ideologues in high places) in the fact that such evidence is no longer with us.

Gravity is the most major source of stress on land animals. The same attenuation of gravity that allowed the super animals of past ages to exist was almost certainly the major factor in the antediluvian lifespans that we read of in Genesis and other antique literature. A society in which the best minds and talent had two or three centuries in which to work could advance more rapidly than human society has within recorded history.

Splash Saltations

The view of the origin of modern man on this planet that we are proposing involves man being brought to this planet rather than originating here. Aside from the question of how humans got from one planet to another in ancient times, there is a more general question of how large numbers of animal species arrived on this planet at a number of distinct points during past ages. In this work we refer to such an arrival as a 'saltation,' a term borrowed from the geological sciences describing a specific type of particle transport by fluids such as wind or water. It occurs when loose material is removed and deposited in a different location — in our case this removal to a different location would involve interplanetary transportations, a not entirely outlandish concept in light of the fact that we know there have been interchanges of materials between Mars and Earth in the form of the Shergotty meteorites.[1]

At least some of these saltation events could be described as resembling a play or opera in which a curtain comes down ending a scene, and then rises to reveal an entirely new cast of characters. There is a related question as to the cause of the fantastic muck deposits found in Canada, Alaska, and Siberia.

There is also the evidence, as we note in one of the appendices to this work, of a spacefaring civilization on Mars in prehistoric times, and it is not inconceivable that organic life arrived via that avenue. However, aside from that possibility, there is also evidence of what one might term "splash saltations" taking place in prehistoric times.

The cover of the December 4, 1995 issue of Time Magazine showed an anomalocaris under the heading, "Evolution's Big Bang":

---

[1] See: Wikipedia entry for "Shergotty meteorite," http://en.wikipedia.org/wiki/Shergotty_meteorite

Time was apparently working on producing an ultimate oxymoron ("evolution's Big Bang"). What they were talking about was the Cambrian Explosion, or the start point for complex life forms on this planet. It is said that if the history of this planet were represented as a 24-hour clock, then in all the time from zero to 2100 hours, the planet saw only sponges and other very simple life forms. And then, in a space of time too short to do anything other than guess about and that is often given as something like ten million years, most or all of the present animal phyla, or basic kinds, appeared .

Clearly there no version of evolution that would account for anything remotely like a 'Big Bang' in the appearance of new organisms.

One thing that COULD account for something like the Cambrian explosion is what we call 'splash saltations'. What that would amount to would be another planet that already had complex life forms getting too close to Earth, and water and creatures on it being ripped off and then landing in water on this planet, and either taking up residence in that water or walking, crawling, or slithering ashore.

Obviously in our solar system as it presently exists, there are no close interplanetary fly-by situations even remotely suggesting anything like the above scenario. However, in the context of an ancient Saturnian system as a polar aligned configuration of planets, as has already been discussed, such a concept is entirely possible. A bit of a description of that ancient system is in order here before we return to the question of how exactly a "splash saltation" might work.

We know from ancient iconography that Mars and Earth were held together in closer proximity within fairly recent times and it is at least possible that other bodies may have been similarly close to us further back in time. The ancient Saturn, Venus, Mars, Earth alignment when viewed from Earth about 7,000 to 10,000 years ago would have looked something like the following image:

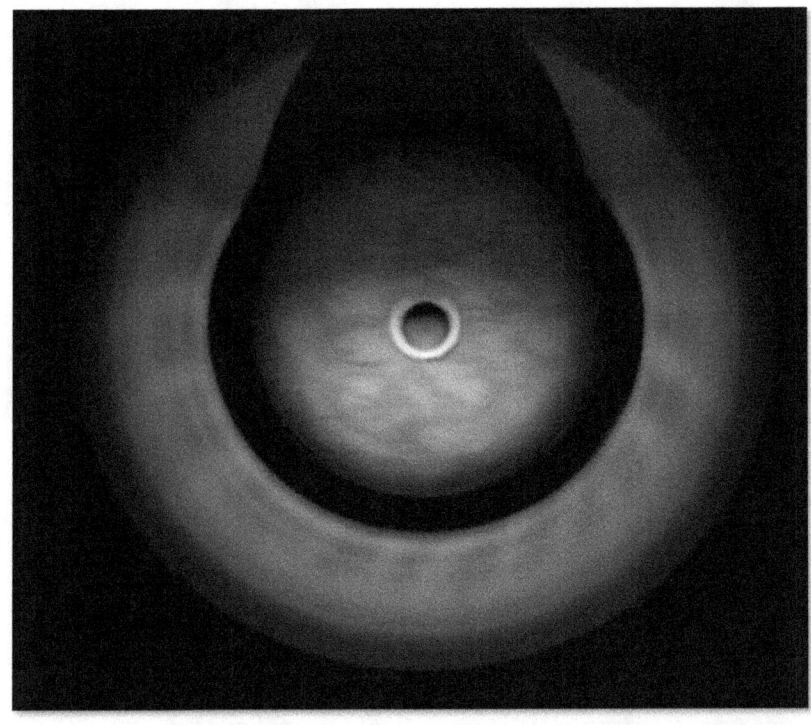

The electrically passive grand polar-aligned conjunction of Saturn, Venus and Mars as seen from Earth at midnight during the Golden Age era. Artist's impression by the authors.

Ancient peoples living around the Mediterranean basin have represented this close polar configuration of planets from our northern skies fairly accurately in a host of glyphs and statues left for posterity. The image above depicts the position of the planet Venus behind Mars where it had settled for a time after its destructive ejection from Saturn's interior. Being a young ejected piece of Saturn's core, the violently hot Venus would have occasionally flared to create the dramatic eight-pointed star burst motif that is associated with this configuration (see below).

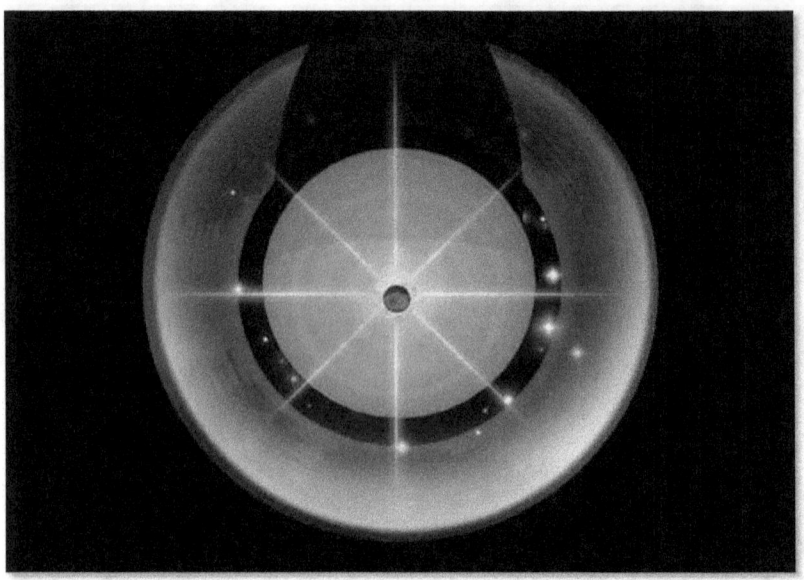

An artist's impression of an electrically active Saturn/Venus/Mars polar conjunction as seen in ancient times. Venus is flaring behind Mars with both smaller planets silhouetted by Saturn. Saturn's rings form the distinctive crescent/bull horns shape due to the Sun's light striking Saturn at a 27° angle and casting a shadow on its rings.

Saturn's ice crystal rings, fashioned from water that was equatorially ejected during Saturn's flare-up after its cataclysmic capture by the Sun, complete the bull horns/crescent that often framed this ancient configuration. The Sun's light striking Saturn at a 27° angle would have cast a shadow on the rings to create this distinctive 'bull horns' or so-called 'moon crescent' shape that is so often depicted in ancient motifs as encapsulating a glowing star (Saturn and Venus). The Shamesh glyph (see image below) is the basis for the familiar Islamic symbol (also seen below,) which is normally assumed to represent a star inside the crescent moon. Once again it only requires a tiny bit of thought to grasp the reality that if an actual main sequence star like our sun ever got between us and the moon, we would be fried. . . The above configuration is not only a logical explanation for how these symbols came into being, but also provides us with a perfectly plausible natural explanation for mythical symbols in keeping with the Herbig-Haro object theory for planetary formation (see chapter 'Mankind's Purple Dawn'). It also tells us that at least three other planets were in very close proximity to Earth at one time.

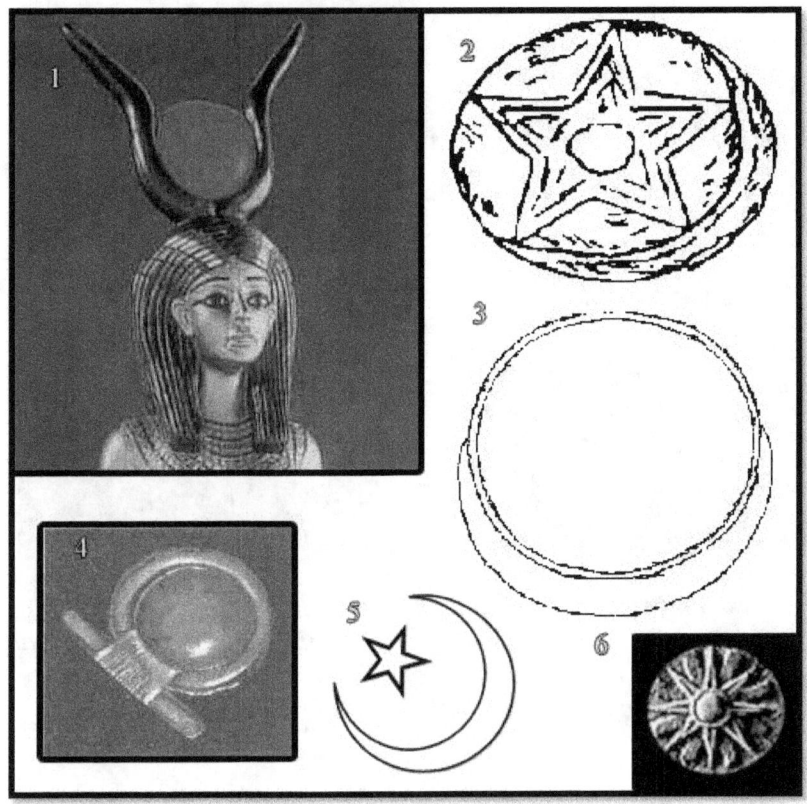

1. An Egyptian depiction of Isis (Venus) showing the goddess' association with the Saturn configuration. 2. Babylonian Shamesh glyph. 3. Egyptian Crescent enclosure. 4. Shen-Bond clasp. 5. Islamic crescent and star symbol. 6. Babylonian Ishtar(Venus) glyph. All images are clear representations of a polar alignment of the planets Saturn, Venus and Mars.

Aside from seeing other planets and a sub brown dwarf star to the north, ancient people also saw a Birkeland current and an electromagnetic flux tube, which they typically viewed as a cosmic mountain; thus Zeus and his associated pantheon were said to live on Mount Olympus while Cronos/Saturn and the Titan gods were said to dwell on Mount Othrys .

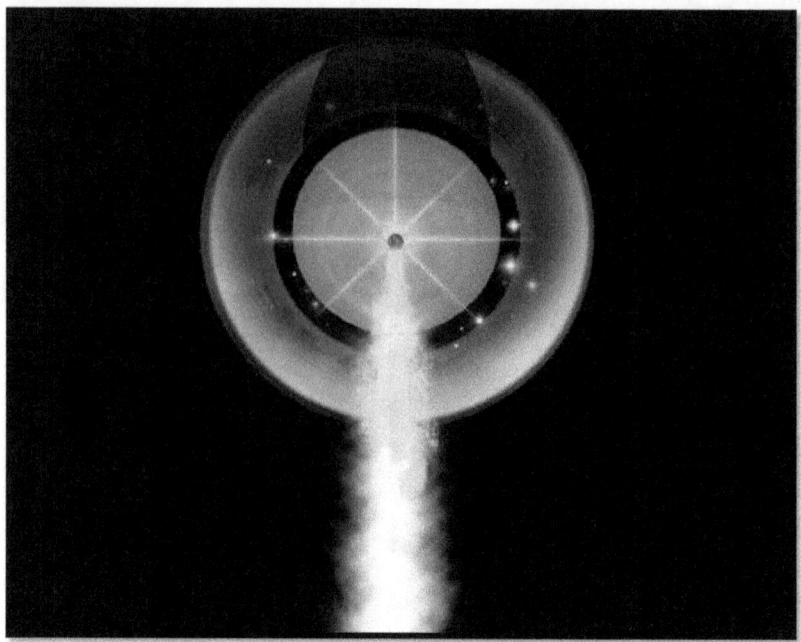

A Birkeland current forming a plasma tornado between Earth and the polar-aligned planets Mars/Venus/Saturn. It was this sight that prompted the mythical stories of a ladder to heaven and a great cosmic mountain between the abode of the gods and Earth.

A similar connection exists today in the form of a Birkeland current seen between Jupiter and its moon Io. From Wikipedia:

> Jupiter's magnetic field lines, which Io crosses, couples Io's atmosphere and neutral cloud to Jupiter's polar upper atmosphere through the generation of an electric current known as the Io flux tube. This current produces an auroral glow in Jupiter's polar regions known as the Io footprint, as well as aurorae in Io's atmosphere. Particles from this auroral interaction act to darken the Jovian polar regions at visible wavelengths. The location of Io and its auroral footprint with respect to the Earth and Jupiter has a

strong influence on Jovian <u>radio</u> emissions from our vantage point:[2]

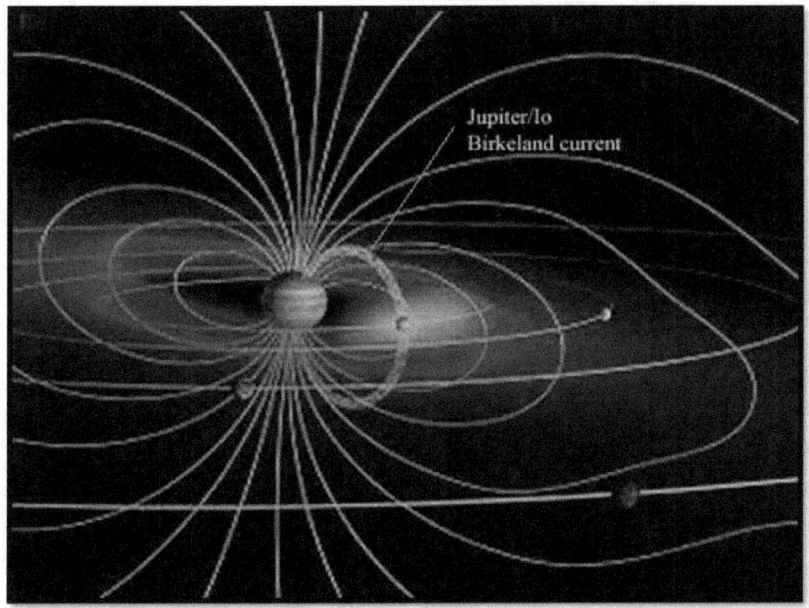

Schematic of Jupiter's magnetosphere and the components influenced by Io (near the center of the image): the plasma torus (in red), the neutral cloud (in yellow), the flux tube (in green), and magnetic field lines (in blue).

A similar Birkeland current stretching between Earth and Saturn would have been highly visible from Earth. Egyptian writings refer to it as Shu, or the pillar of Shu, and describe it as supporting the celestial abode of the gods. Hieroglyphs and other Egyptian art show this phenomenon in a number of ways, however all such amount to a pillar of some sort supporting an enclosed star representing Saturn and possibly its ring system. Any sort of a glyph meaning to illuminate or lighten showed a star with the ring around it supported by lightning forks that pointed upwards, that is, by our friend the Birkeland current.

---

[2] http://en.wikipedia.org/wiki/Io_(moon)

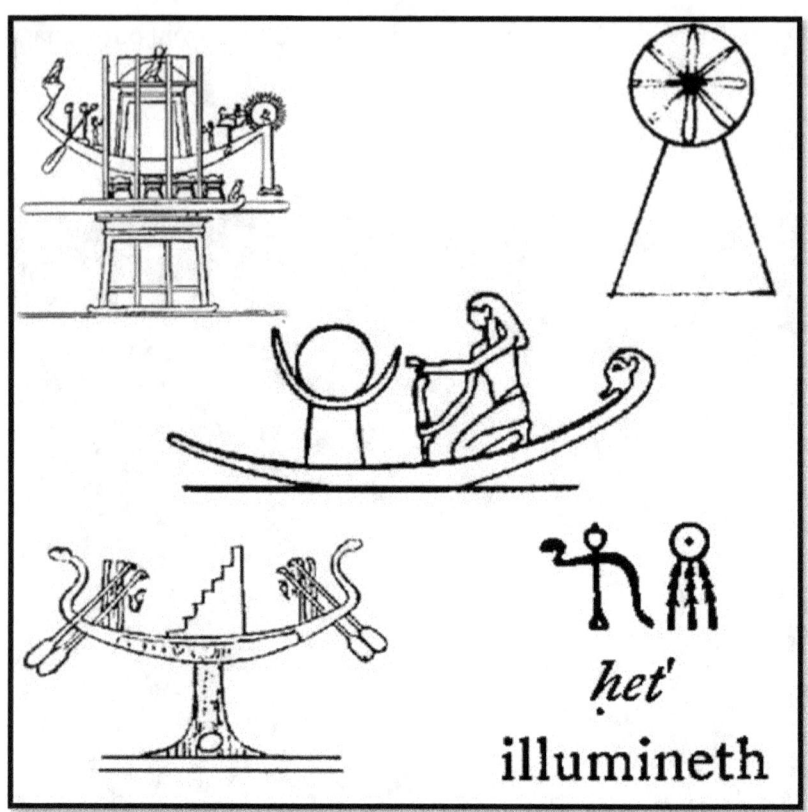

het'

illumineth

The enclosed star was sometimes represented as a human eye and the Birkeland current as a pyramid supporting it. Interestingly enough, some of these depictions, particularly hieroglyphs which signify illumination or brightening, show a pyramid of lightning forks which point **UPWARDS**. All depictions of lightning within historical times show it pointing downwards. The most modern manifestation of this pyramid/eye sort of Saturnian iconography can be seen on the US dollar bill (see image below).

Unlike the auroras and other plasma physics phenomena that we observe today, this ancient Birkeland current, which Greeks called Othrys and the Egyptians called the pillar of Shu, was more than a spectacular light show. Proponents of the electric universe theory view Birkeland currents and the z-pinch effect associated with them as the basic mechanism used by the universe for agglomerating material from plasma form into solid objects including stars, planets, galaxies, and strings of galaxies.[3] Such currents potentially have the power to suck up material like a vacuum cleaner. Consider for a moment the gigantic muck deposits, which are found in

---

[3] E.g., http://www.holoscience.com/wp/the-true-state-of-the-universe/ "Stars cannot be formed by gas and dust falling inward. Rotational energy will put a stop to it. Molecular clouds, from which stars are supposed to form, show no signs of collapsing. And it has never been shown how planets can form from a dusty disk encircling a star. Instead, stars are formed by the strong electromagnetic 'pinch effect,' like beads on a string, along a discharge in a dusty plasma. Each string is a cosmic Birkeland current, which scavenges gas and dust at long range with its 1/r electromagnetic 'pinch' force. Unlike gravity, the electric force can strongly repel as well as attract. So there is no mystery about 'stuff streaming outward' from stars. Nor is it surprising that the streams take the form of thin jets. A thin jet is the hallmark of an electric discharge. Planets, too, are formed not by separate accretion but by electrical expulsion of core material from larger bodies."

Canada, Alaska, and Siberia, the very place where a Birkeland current connecting celestially north Saturn to Earth would have landed:

"In even the most prejudiced murder trial there is one essential element: there has to have been a killing. Fancy legal terminology generally requires a body, the corpus delictus as the TV detective shows are fond of telling us. It would seem reasonable, if one was to promulgate a theory of blitzkrieg slaughter as have Martin and Diamond, to identify where the bodies are buried and then take the reader on a gut-wrenching tour through a graveyard of waste and butchery. We are deprived of this vicarious thrill because the evidence of the destruction of the megafauna suggests a scenario well outside the orthodox interpretation of benign natural processes. Therefore mere mention of the reality of the situation is anathema to most scholars...

...When we inquire if the Alaskan area has similar deposits, we learn that the situation is the same. Early gold miners in Alaska discovered that in many cases they had to strip off a strange deposit popularly called "muck" in order to get to the gold-bearing gravels. The muck was simply a frozen conglomerate of trees and plants, sand and gravels, some volcanic ash, and thousands if not millions of bits of broken bones representing a wide variety of late Pleistocene and modern animals and plants.

Two scholars describe the scenes of destruction and chaos which the muck represents. Frank Hibben, in an article surveying the evidence of early man in Alaska, said that while the formation of muck was not clear,". . . there is ample evidence that at least portions of this material were deposited under catastrophic conditions. Mammal remains are for the most part dismembered and dis-articulated, even though some fragments yet retain in this frozen state, portions of ligaments, skin, hair, and flesh. Twisted and torn trees are piled in splintered masses concentrated in what must be regarded as ephemeral canyons or arroyo cuts."'1

Stephen Taber's report echoes the same conditions. He says: "Fossil bones are astonishingly abundant in frozen ground of Alaska, but articulated bones are scarce, and complete skeletons, except for rodents that died in their burrows, are almost un- known."[2] Many laypeople will be confused by this technical language and fail to grasp what Taber is saying, allowing him to imply a benign orthodox interpretation when the situation re- quires that a clearer picture be drawn.

When a scholar says "articulation" of bones he means an arrangement of bones that a person observing them would identify as a complete skeleton and from which an experienced observer could identify the species. To say that articulated bones are scarce, then, means that the bones are scattered and mixed so badly that expert examination is needed to identify even the bone itself, let alone the species from which it comes. Remember this problem of articulation, for we shall meet it again in another context. Taber concludes with the observation that "the dispersal of the bones is as striking as their abundance and indicates general destruction of soft parts prior to burial."[13] In other words, Alaskan muck is a gigantic pile of bones representing a bewildering number of species, a good number of them the megafauna I have been discussing.

We find the missing megafauna of the late Pleistocene in the Siberian islands, in the islands north of Alaska, and in the muck in the Alaskan interior. Obviously we have here victims of an immense catastrophe which swept continents and left the debris in the far northern latitudes piled in jumbled masses that now form decent-sized islands. Most anthropologists and archaeologists avoid discussing these deposits because the orthodox uniformitarian interpretation of the natural processes precludes sudden unpredictable actions..."[4]

These muck deposits are basically the remains of an obliterated world and the depths of these jumbles of animal and vegetable remains of that world are so great that even something like the flood at the time of Noah may not suffice as a cause. Dwardu Cardona has suggested that so long as the ancient Birkeland current remained anchored at the North Pole it was at least not dangerous; but that at some time during or after the breakup of the Saturnian system, the thing lost its mooring and, like a vacuum cleaner, simply sucked up

[4] Vine Deloria "Red Earth, White Lies" page 114

all of this material and then dumped it in the heaps that we call the Alaskan, Canadian, and Siberian muck.

Such as the evidence for the activity of a Birkeland current vacuuming up and re-depositing material on our own planet in the recent past. Getting back to the point we were trying to make prior to the long detour, this is the most realistic version of what a splash saltation might amount to. One can picture such a current between two planets carrying living creatures from one to the other, particularly if water were also involved in the process — and there is no shortage of water where brown dwarf stars are concerned. The creatures involved in such an event would basically be passing through an electrically active plasma straw.

The more straightforward notion of planetary bodies simply getting too close together and creatures from the one landing on the other simply due to the effect of gravity, is much more problematical. It might be possible, but we would not want to bet money on the notion that it had ever occurred.

A primordial 'splash saltation' powered by a Birkeland current stretching between proto-Saturn and Earth. The view is over the Arctic Sea looking north from the Siberian arctic coast. The Birkeland current creates a huge hurricane-like vortex at the North Pole both sucking up and releasing organisms, space-borne bacteria and cosmic dust.

156

The other possibility for transferring creatures from one planet to another in our antique system requires less explanation. We note in an appendix that there was a space-faring civilization based on Mars in prehistoric times. The inhabitants of Mars at that time would have been aware of the forthcoming capture of the Saturnian system by our present sun and its associated mayhem and they would not have known if anything in either system would remain habitable. It is a reasonable assumption that they left either for the near stars or for deep space. They may have been able to ascertain that Earth alone had any chance of remaining habitable and, if so, like some distantly ancient space-faring version of Noah's Ark, they may have gathered creatures from any other inhabited worlds, either in the Saturnian system or in the ancient system of our present sun, and brought them here to give them the best possible chance of survival, *IF* Earth survived. . .

Summary and Takeaways from this Chapter

The Cambrian explosion and several other events, which are detectable from the fossil record, amounted to very large changes in our living world that took place suddenly.

An attempt to catalog the ways by which various kinds of creatures might have arrived on earth would start with the obvious possibilities of them being created here or brought here by a party or parties capable of that. Beyond that however, there are a couple of possibilities that most readers will not have seen either in religious or scientific literature. These possibilities involved Birkeland currents and the power that Birkeland currents have to agglomerate or actually vacuum up solid materials.

The gigantic muck deposits found in Canada, Alaska, and Siberia amount to the remains of an obliterated world and it is not clear that even so great a clause as the flood at the time of Noah could have caused such a thing. One startling possibility that has been proposed is that the primordial Birkeland current associated with the ancient Saturnian system may have simply sucked up that gigantic mass of material and then deposited it in the multi-thousand foot layers that are found.

There is a similar possibility of living creatures from one body being transported to another body by a Birkeland current stretching out between them. Assuming that water was also involved, this experience may have been like passing through a straw, to land in an alien world.

Part II: The Ganymede Hypothesis

The Antique Solar System

The Sun and its Planets Before the Arrival of Saturn

The capture of an interloping system of Saturnian planets[1] by the Sun begs the question as to what the Solar System looked like before Saturn's calamitous arrival. Just how would our current Solar System have once appeared sans the planets Venus, Earth, Mars, Saturn, Uranus, and Neptune?

The pithy answer would be that only Mercury and Jupiter, and possibly Pluto, would remain as the Sun's original native collection of planets. There is also a possibility of a now shattered remnant of a planet whose debris currently makes up the asteroid belt.[2] However, that would not fully answer the fundamental question as to what the arrangement and orbital characteristics of these original

---

[1] In this work Saturn Theory is largely defined as the concept of the planet Saturn originally being a lone extra-solar free-floating body accompanied by a number of satellites, generally positing at minimum two known satellites of the system: Mars and Earth. The emergence of Venus was a later effect of the system's catastrophic entry into the Sun's electrical sphere of influence. However, there are substantive grounds to also include the gas giants Uranus and Neptune as originally part of what is called the Saturnian System (see chapter titled "DISCOVERING THE ANTIQUE SOLAR SYSTEM"). In such a scenario, Uranus and Neptune would have remained hidden behind Saturn from the Earth's perspective, thus accounting for little, if any, mythological reference to these two bodies regards earthbound observations of the primordial proto-Saturn sphere; the latter object designated by comparative world mythology to have sat primarily at Earth's celestial north for the duration of such a configuration.
Given a 'polar' configuration, or axial alignment (as demanded by mythology), for this Saturnian line-up of planets, we have the following possible north-to-south axial arrangement: **Uranus/Neptune/Saturn(sub-brown dwarf)/Mars/Earth**. The latter addition of **Venus** would place that newly formed planet between Saturn and Mars sometime after the Saturnian system's capture by the Sun and its eventual breakup.

[2] Scientists in the 20th century called this hypothetical planet Phaeton.

planets would have been in relation to the Sun. Given the circumstances, it's fair to assume that these original planets would not necessarily have been in the positions they currently occupy. Their orbits, and even their physical appearance, could be expected to have been noticeably different to what is observed today.

How so?

For a more in-depth answer to this question, we can turn to some of the recent discoveries involving extra-solar planets, or exoplanets as they are called, found to be circling other sun-like stars,[3] particularly those planets found to be orbiting their parent star in the so-called 'habitable zone.' The results have generally been seen to clash with most mainstream theories of planetary formation, particularly the Solar Nebular Disk Model (SNDM). That model requires planetary systems to conform to accepted gravitational accretion theories and reflect the general orbital sequences and arrangements of the various planetary types we see in our own solar system.

When the discovery of exoplanets was first confirmed,[4] it was natural that there would be a sampling bias towards finding large, gas-like giants since their enormous physical size made them easier to detect than smaller sized terrestrial-type planets like Earth. What was not anticipated, however, was that these huge gas giants, some many times the mass of Jupiter, would be found circling in extremely close orbits around their host stars. This fact is completely at odds with the accepted SNDM notion of planetary formation. The gravitational-based nebular model for the formation

---

[3] A 'sun-like' star is currently assumed/defined as a main-sequence star of spectral classes late-F, G, or early-K and is without a close stellar companion; see G. Marcy; Butler, R. Paul; Fischer, Debra; Vogt, Steven; Wright, Jason T.; Tinney, Chris G.; Jones, Hugh R. A. *Observed Properties of Exoplanets: Masses, Orbits and Metallicities*, (2005), see arXiv:astro-ph/0505003

[4] Canadian astronomers Bruce Campbell, G. A. H. Walker, and Stephenson Yang announced in 1988 the first discovery of an exoplanet orbiting the star *Gamma Cephei. This* was subsequently fully confirmed in 2003. The 1992 discovery of two planets circling the pulsar *PSR 1257+12* is generally recognized as the first fully confirmed discovery of an extra-solar planetary system.

of a solar system calls for gas-giant planets to inhabit the system's outer edges, while smaller terrestrial planets are expected to take up the innermost orbits. When Jupiter-sized planets, or 'hot Jupiters' as they were subsequently called, started appearing where Earth-like planets should be, journalists covering mainstream science started talking about going back to the drawing board:

> "As astronomers discovered the first extra-solar planets, it quickly became obvious that the formation theories that we'd built on our own Solar System were only part of the story. They didn't predict the vast number of hot Jupiters astronomers found nearly everywhere. Astronomers went back to the drawing board *to put more details into the theory,* . . ."[5] (Emphasis ours)

The above quote highlights the tendency of mainstream researchers to try to force-fit new and inconvenient data into increasingly untenable theories (*"put more details into the theory"*). This illustrates the weakness of the accepted SNDM paradigm in the area of prediction. An American textbook on planetary science goes even further in stating:

- Nebular theory predicts that massive Jupiter-like planets should not form inside the frost line (at << 5 AU)
- Discovery of "hot Jupiters" has forced re-examination of nebular theory [6]

---

[5] Jon Voisey, *Rocky, Low-Mass Planet Discovered by Microlensing,* June 17, 2011, Universe Today.com; http://www.universetoday.com/86841/rocky-low-mass-planet-discovered-by-microlensing/

[6] Bennett, Donahue, Schneider, and Voit; *The Cosmic Perspective,* 5th Edition, Chapter 13.3, pdf. See: http://burro.astr.cwru.edu/Academics/Astr201/

With gas-giant exoplanets now regularly discovered within $1 - 2$ AU[7] of their host stars, we now have growing data pointing to the common existence of gas-giants within the habitable zone of far-off distant sun-like stars - gas-giants not unlike our own Jupiter. It would not then be untoward to suggest that, given the then absence of the other major planets seen today, the Sun's sole and original gas-giant, namely Jupiter, would most likely have been found in a much closer orbit to the one it currently holds, as seems to be the norm elsewhere. If such an orbit were within $1 - 2$ AU of the Sun (where Earth is now), then it is reasonable to also suggest that Jupiter's three currently ice covered moons, Europa, Ganymede, and Callisto, would have enjoyed liquid water environments spectacularly conducive to the existence of aquatic-based life had they also borne substantive atmospheres.

But we get a little ahead of ourselves. . .

Jupiter and its Galilean moons. All four moons are candidates in

---

[7] 1 AU = 1 Astronomical Unit, i.e. the distance of Earth from the Sun; 149,597,870,700 meters (92,955,807.273 miles).

the current search for extraterrestrial life forms. Image courtesy of NASA.

So, an antique solar system consisting of the Sun and only one gas-giant, the planet Jupiter and its moons, and the tiny, almost inconsequential planets Mercury and Pluto, begins to look remarkably similar to the systems being observed around other far-off stars. Taking this into account, this would conceivably place Mercury in a close, moon-like relationship with the Sun. Jupiter would possibly display aspects in keeping with a former sub-brown dwarf star (more on this later) while enjoying a binary relationship with its main-sequence yellow dwarf host star. That Jupiter today shows enough signs of being a failed sub-brown dwarf are added indications pointing to that planet's possible former life as a binary sub-stellar companion to the Sun.

The idea that Jupiter may have once been a star-like object in its own right is not new. As the largest object in our solar system after the Sun, Jupiter continues to defy standard models attempting to account for its formation from an inert nebular disk of dust and gas. Observations of Jupiter have established that it sheds more heat than it absorbs and that it is the most electrically active planet in the Solar System — features to be expected more in a star-like object, and another unexpected contradiction to the accepted SNDM theory of planetary formation. The giant planet operates like a mini-solar system in its own right, with sixty-seven confirmed satellites. The largest of these is Ganymede, a moon bigger than Mercury and almost three-quarters as large in volume as Mars.

Then there are also the more direct electrical elements proposed by the Electric Universe (EU) model to be factored into the Jupiter equation.[8] It has the largest magnetosphere (initially an electrically-

---

[8] It should be born in mind that the Electric Universe (EU) model basically dismisses gravity as the single most important force in the universe and recognises the presence of electrical currents in space (plasma cosmology) as the foremost force in shaping the observable universe – including stars. Under this paradigm, all stars, including brown dwarfs, are an electrical discharge phenomenon in space conforming to basic electrical principles,

created property according to EU principles) in the Solar System and, despite its massive size, rotates faster than any other planet (10 hours), a feature not unlike the much more enormous Sun, which rotates fully within the relatively speedy period of 25.6 days. The relatively recent discovery, that Jupiter has its own set of rings, has prompted the conclusion that they are electrically charged.[9] It has also been long known that Jupiter is a copious source of x-rays[10] and radio emissions, all tell-tale signs pointing to Jupiter as former a sub-brown dwarf star as opposed to an electrically inert ball of accreted stellar dust.

Another interesting point in assessing the Sun and Jupiter's relationship in our proposed antique solar system is that Jupiter exhibits a near non-existent axial tilt (only 3.13°), making it the most upright body in the Solar System, second only to the Sun itself, and an indication that both share a common origin. Contrastingly, and as discussed in a previous chapter, of the afore mentioned planets that are postulated to have made up the later interpolating Saturnian system, all but Uranus and Venus exhibit axial tilts between a narrow range of 24 and 27 degrees,[11] again indicating a different origin for these planets compared to Jupiter.

---

and *not* controlled hydrogen fusion reactions as the standard nuclear model of the sun insists. For a review of the EU model's application to stars, see: Wallace Thornhill, *Our Misunderstood Sun,* March 2010, http://www.holoscience.com/wp/our-misunderstood-sun/; Also see Donald E. Scott, *The Electric Sky,* Mikamer Publishing, Portland, Oregon, 2006.

[9] "Jupiter's Shadow Sculpts Its Rings," NASA *news release*, 30 April, 2008. see http://www.nasa.gov/topics/solarsystem/features/galileo-20080430.html

[10] See Gladstone *et al*, *'A pulsating auroral X-ray hot spot on Jupiter,'* Nature magazine, Feb. 28, 2002.

[11] Uranus virtually, and bizarrely, lays spinning on its side with its geographical poles roughly pointed along its orbital plain, the undoubted result of a catastrophically huge disruption to its motion and the possible result of the Saturnian system's initial entry into the Sun's electrical influence. Venus, on the other hand, has a retrograde rotation, making it unique in the Solar System. This retrograde spin cannot plausibly be primordial. It has to have arisen via interaction with another body in the

Having now proposed a possible ancient pre-Saturnian relationship between the Sun and Jupiter based on current observations of extra-solar planetary systems, it remains for us to speculate on the life of Jupiter as a possible former sub-brown dwarf enjoying a close-orbit binary relationship with the Sun. The first question then is: Do such binary relationships exist today between recognized brown/sub-brown dwarf stars and sun-like stars?

While the short answer is 'yes,' a more complete answer turns out to be problematic. This is because, while extra-solar 'hot-Jupiters' and even super-Jupiters are seen to enjoy close-in orbits around their parent stars, *most* confirmed brown dwarfs in binary relationships seemingly prefer a much wider orbit with their sun-like host. I have emphasized the word *most* because, as we shall see shortly, there are always exceptions to the rule. But for the moment, here is the consensus view:

> "Dozens of brown dwarfs have been discovered orbiting normal stars. But astronomers have always wondered about the paucity of close-in brown dwarfs: While many giant planets have been found in small orbits, whirling around their sunlike stars in just a few days, the more massive brown dwarfs appear to shun these intimate relationships."[12]

The article quoted above goes on to explain that a brown dwarf, a more massive sub-stellar object than a gas-giant planet, would probably find itself swallowed up by its parent star if it came in too close. This assumption is based entirely on what is called a 'brown dwarf desert', the currently observed absence of brown dwarfs in close-in orbits around sun-like stars.

---

system and the curious phase-lock between Venus and Earth (showing observers on Earth the same face at inferior conjunctions) indicates that the body in question was Earth.

[12] Govert Schilling, *Brown Dwarfs Being Gobbled Up by Parent Stars*, Science Now, 14 June, 2012, see: http://news.sciencemag.org/sciencenow/2012/06/brown-dwarfs-being-gobbled-up-by.html

"Planetary scientist Tristan Guillot of the Côte d'Azur Observatory in Nice, France, has an explanation for this "brown dwarf desert." At the meeting [220th meeting of the American Astronomical Society], he argued that brown dwarfs in tight orbits get devoured by their sunlike parent stars. "Some fall into their stars very quickly," he says."[13]

And yet, we have a plethora of Jupiter-sized and larger bodies enjoying close-in orbits without them seemingly in danger of imminently falling into their parent star. With brown dwarfs and Jupiter-sized planets being almost indistinguishable in volume, would it not be likely that the currently observed close-in 'hot Jupiters' may actually be the *less massive* remnants of full-blown brown dwarfs that have drifted in closer?

Furthermore to this question, are there any examples of a *brown dwarf* orbiting its parent star in what could be termed a 'close-in' orbit despite the so-called 'brown dwarf desert'? Of course, the close-in orbits of hot Jupiters are usually indicative of orbits closer than 1 AU, hence their extremely high surface temperatures due to their close proximity to their star. Ideally, for the sake of the case we are building here, it would be sufficient to find a brown dwarf in the habitable zone of 1 – 2 AU distance from its star. So, is there such an object?

The Strange Case of 'HD 141937 b'

As usual, there is always the exception to the rule. In the case of the supposed 'brown dwarf desert' as discussed above, we may not have a recognized brown dwarf situated closer than 1 AU to its host star, but we do have one enticingly located within the habitable zone of its host star. That object is the suspected *sub*-brown dwarf designated *HD 141937 b*, a currently designated exoplanet with a 9.7 times mass of Jupiter circling a star 109 light-years away in the constellation of Libra.

---

[13] Ibid.

However, *HD 141937 b* is most likely a brown dwarf because its currently calculated mass is its estimated minimum and any subsequent upward revision to its mass would place it firmly in the category of a low-mass brown dwarf.[14] But more importantly for our purposes, it is *HD 141937 b*'s eccentric orbit that is of most interest. In this case,we have a suspected brown dwarf placed well inside its host star's habitable zone for a significant part of its orbit. The following diagram demonstrates this, with our own sun's planets plotted with *HD 141937 b* for comparison:

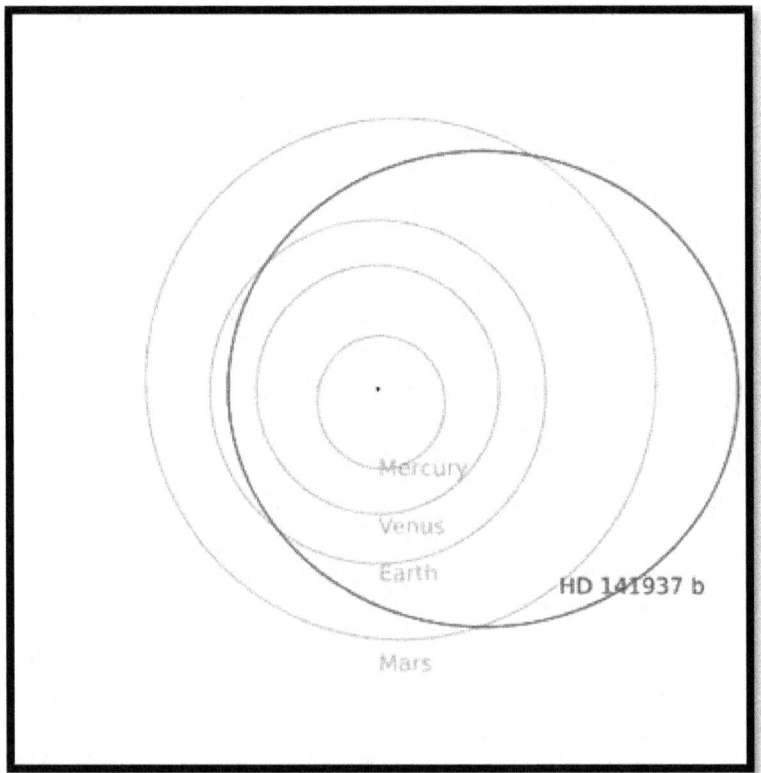

Suspected sub-brown dwarf HD 141937 b orbit comparison with our own Solar System. The habitable zone is roughly between Earth

[14] See*: The CORALIE survey for southern extra-solar planets,* A&A 390, 267-279 (2002) DOI: 10.1051/0004-6361:20020685, Copyright ESO 2002 Published by EDP Sciences, http://www.aanda.org/index.php?option=com_article&access=standard&Itemid=129&url=/articles/aa/full/2002/28/aa2416/aa2416.html

and just outside Mars' orbit. Courtesy of the Visual Exoplanet Catalogue.

If *HD 141937 b* is a *sub*-brown dwarf, then there is the possibility it has its own terrestrial-type satellites, or moons, that would be sufficiently warmed for habitation by the low temperature glow of *HD 141937 b* itself during the small period of time it was outside its host star's habitable zone. This would make life on any of its satellites a possibility despite the bizarre and extreme seasonal variations they would encounter orbiting their sub-brown dwarf host as they circled the system's main star.

We have now come to the point where we need to discuss what exactly a brown dwarf or a *sub*-brown dwarf is, and just how likely it is that Jupiter may have once been such an object. It is important to establish this understanding because determining what type of object Jupiter might have been in our postulated Antique Solar System will also determine what its relationship to the Sun was and, not least of all, any possible life-giving effect Jupiter may have had on its native moons.

Ancient Jupiter as a Glowing Sub-Brown Dwarf Star

The confirmation of the existence of brown dwarf stars is almost as recent in the history of astronomy as the confirmation of exoplanets.[15] Now they are thought to be more plentiful than regular sun-like stars, but still very difficult to detect from Earth with optical technologies.

One of the difficulties experienced by astronomers in differentiating a brown dwarf from a gas giant planet is that brown dwarfs have roughly the same radius as Jupiter. It is therefore not a major

---

[15] http://en.wikipedia.org/wiki/T-dwarf#History, Brown dwarfs, a term coined by Jill Tarter in 1975, were originally called black dwarfs, a classification for dark substellar objects floating freely in space that were too low in mass to sustain hydrogen fusion.

surprise that confirmation of the existence of brown dwarfs and exoplanets occurred almost simultaneously.

> "The term "brown dwarf" was coined in 1975 by Jill Tarter (1975) to describe substellar-mass objects (SMOs), but astronomers had to wait 20 years before the announcement of the discovery of the first unimpeachable example, Gliese 229B (Oppenheimer et al. 1995). That same day, the first extrasolar giant planet (EGP1) was announced (Mayor and Queloz 1995) and it *startled the world by being 100 times closer to its primary than Jupiter is to the Sun.*"[16] (Emphasis ours)

(Again, we see reference to the startling and unexpected discrepancies with accepted mainstream planetary theory that is provoked whenever a gas-giant planet is found close to its host star.)

Amongst brown dwarf stars there is only, on average, a 10 – 15% difference in their radii, making their volume virtually indistinguishable from gas-giants like Jupiter. However, what marks the difference between a gas-giant planet and a brown dwarf is the *mass* of the object in question. In conventional thinking, a sub-stellar object needs to have at least thirteen times more *mass*[17] than Jupiter to qualify as a brown dwarf. An object with a *mass* lower than thirteen Jupiters, yet more *massive* than Jupiter itself, can be said to be a *sub*-brown dwarf.

---

[16] Adam Burrows, W.B. Hubbard, J.I. Lunine, James Liebert, "The Theory of Brown Dwarfs and Extrasolar Giant Planets," 2001, arXiv:astro-ph/0103383v1 , PDF, page 6

[17] In scientific use, the term *mass* should never be confused with weight or the amount or density of matter. From Wikipedia: In physics, mass (from Greek μᾶζα "barley cake, lump (of dough)"), more specifically inertial mass, is a quantitative measure of an object's resistance to acceleration. In addition to this, gravitational mass is a measure of magnitude of the gravitational force, which is exerted by an object (active gravitational mass), or experienced by an object (passive gravitational force) when interacting with a second object.

With *mass* then being a prime determining factor in identifying a brown dwarf, mainstream thinking then turns to the presence of lithium as another deciding element. It is the presence of lithium that helps scientists separate low-mass brown dwarfs from gas-giants and any high-mass brown dwarfs from low-mass stars[18]. Because it is believed that true stellar objects, like our sun, run on nuclear fusion, the heat within such stellar objects apparently burns up and gets rid of lithium quickly. This means, then, that the continuing presence of lithium in a sub-stellar object more massive than Jupiter should indicate that it is a type of brown dwarf.

> "Though SMOs [substellar mass objects] are characterized by the fact that they don't generate sufficient power by thermonuclear processes to balance their surface radiative losses, they can have thermonuclear phases, however partial or temporary. Objects more massive than ~13 MJ [Mass of Jupiter] will burn deuterium via the $p+d \rightarrow \gamma + {}^{3}He$ reaction and objects more massive than ~0.06 $M_{\odot}$ (~63 M_J) will burn lithium isotopes via the $p + {}^{7}Li \rightarrow 2\alpha$ and $p + {}^{6}Li \rightarrow \alpha + {}^{3}He$ reactions."[19] *(Authors' note - the equations are included purely for academic accuracy.)*

This brings us to the existence of low-end, or *low-mass* brown dwarfs, a sub-class of sub-stellar objects thought to be incapable of initiating the kind of fusion that can burn lithium quickly, or even fuse deuterium for that matter. These are the class of objects referred to as *sub*-brown dwarf stars. Differentiating these sub-types of brown dwarfs from a gas-giant planet is particularly difficult.

There are, of course, other supposed tell-tale signs distinguishing a sub-stellar object as a brown dwarf from a gas-giant planet. X-ray and near-infrared spectral analysis can contribute to detecting the low heat glow associated with brown dwarfs — a heat signature that

---

[18] http://en.wikipedia.org/wiki/Brown_dwarf
[19] Adam Burrows, W.B. Hubbard, J.I. Lunine, James Liebert, "The Theory of Brown Dwarfs and Extrasolar Giant Planets," 2001, arXiv:astro-ph/0103383v1 , PDF, page 6

is typically too low to make the object a low-mass star and too high for it to be a planet. Depending on how much a brown dwarf glows and the composition of whatever is supposedly feeding that glow, a brown dwarf will be categorized either as L-class, T-class or Y-class object. If Jupiter had once been a brown dwarf, it would most likely have been a T-class brown dwarf, simply based on the present chemical composition of Jupiter and its large moons as per Wikipedia's "brown dwarf" entry on T-class dwarfs, particularly concerning sodium, potassium, and methane.

According to a standard rundown on what constitutes a brown dwarf, presented below and most probably originally lifted from Wikipedia's brown dwarf entry, it really all boils down to how much the sub-stellar object in question 'glows' as to whether it is a brown dwarf, sub-brown dwarf, or a gas giant planet:

> "The strongest spectral emission of brown dwarfs is in the infrared and that is how present day astronomers study them. *Old brown dwarfs will accumulate methane in their atmosphere*, a compound often taken to indicate active organic molecule kinetics. Atmospheric *temperatures of brown dwarfs range from 2500K to 750K.*"[20] (*Emphasis ours*)

The above quote is helpful in two ways:

> 1. It identifies methane as a component in an <u>old</u> brown dwarf's atmosphere and;

> 2. It tells us that brown dwarfs should put out heat at a temperature somewhere between 2,500 Kelvin and 750 Kelvin.

---

[20] merlynne6, *Are Jupiter and Saturn Brown Dwarfs, Not Planets?* See: http://www.environmentalgraffiti.com/sciencetech/brown-dwarfs-are-neither-stars-nor-planets/5140#DvI6lX3EvQsGwZ6A.99

Interestingly, Jupiter has methane in its atmosphere — supposedly a trait of an *old* brown dwarf, as noted. We also know that Jupiter, like Saturn, puts out more heat than it receives from the Sun . . . another indicating factor pointing to a star-like past for the current gas-giant. Jupiter also pumps out X-rays and radio emissions and is surrounded by a massively strong magnetosphere. The EU model views this as an indication of high-powered inter-stellar and inter-planetary electrical activity,[21] and not just a by-product of internal convection and gravity. In fact, radiation from Jupiter today is still so intense that it renders the idea of human colonization of Jupiter's moons virtually impossible (except for the moon Ganymede, which has its own protective magnetosphere, but more on that later).

Brown Dwarfs as Heavenly Water Carriers

Yet another intriguing aspect of brown dwarfs is their apparent abundance of water. Some of the cooler brown dwarfs are actually speculated to possess cloudy atmospheres full of water.

> "Central topics of SMO [substellar mass object, i.e. brown dwarfs] theory are atmospheric chemistry, thermo-chemical databases, and the molecular abundances of the major atmospheric constituents. For solar metallicity, near and above brown dwarf/EGP photospheres the dominant equilibrium form of carbon is CH4 or CO, that of oxygen is H_2O, . . . (emphasis ours)
>
> ". . .The corresponding SMO mass below which a H2O cloud can form within a Hubble time is ~30-40 M_J . Hence, we should expect to discover many brown dwarfs capped with H2O clouds. *Such objects ("water cloud" dwarfs)*

---

[21] Refer back to footnote 8 of this chapter.

would constitute another spectroscopic class after the T dwarfs."[22] (Emphasis ours)

Water, water everywhere it would seem! As noted earlier, the type of brown dwarf that Jupiter may have been corresponds to the spectroscopic T-class of brown dwarf, or lower — a very wet beast indeed. If Jupiter was once a *sub*-brown dwarf with a copious cloud of water drifting around it, then maybe this is an indication as to where all the ice came from that currently covers three of its four major moons. It would be entirely plausible for Jovian water to have misted about inside Jupiter's magnetosphere and down onto its satellites during any speculated phase in which Jupiter might have been a *sub*-brown dwarf.

So, is there any sign of water in Jupiter's atmosphere today?

"There may also be a thin layer of water clouds underlying the ammonia layer [of Jupiter's atmosphere], as evidenced by flashes of lightning detected in the atmosphere of Jupiter. This is caused by water's polarity, which makes it capable of creating the charge separation needed to produce lightning. These electrical discharges can be up to a thousand times as powerful as lightning on the Earth. The water clouds can form thunderstorms driven by the heat rising from the interior."[23]

---

[22] Adam Burrows, W.B. Hubbard, J.I. Lunine, James Liebert, "The Theory of Brown Dwarfs and Extrasolar Giant Planets," 2001, arXiv:astro-ph/0103383v1 , PDF, pages 6 – 7.

[23] Wikipedia, "*Jupiter*", see http://en.wikipedia.org/wiki/Jupiter#Atmosphere. References employed for quoted exert: Elkins-Tanton, Linda T. (2006). *Jupiter and Saturn*. New York: Chelsea House., Watanabe, Susan, ed. (February 25, 2006). "Surprising Jupiter: Busy Galileo spacecraft showed jovian system is full of surprises". NASA. http://www.nasa.gov/vision/universe/solarsystem/galileo_end.html., and Kerr, Richard A. (2000). "Deep, Moist Heat Drives Jovian Weather". *Science* **287** (5455): 946–947. doi:10.1126/science.287.5455.946b. http://www.sciencemag.org/cgi/content/full/287/5455/946b.

What this means is that Jupiter does currently exhibit enough traits identifying it as having possibly once been *at least* a *sub*-brown dwarf star. There are even grounds to consider it as having once been a possible "water cloud" dwarf, if that classification ever catches on. If this had been so, then it remains for us to determine what Jupiter's heat output may have been and if this may have contributed to habitable conditions on any of its major moons. This is something that will be looked at in more detail in a later section of this chapter.

So, to summarize thus far, what we have learned is that the more massive full-blown brown dwarf stars in binary relationships with sun-like stars are unlikely to be found orbiting closely to their parent stars. . . unless they are suspected *sub*-brown dwarf stars like the afore mentioned *HD 141937 b*. However, what we also know is that *sub*-brown dwarf stars are almost indistinguishable in their volume-size from Jupiter-sized planets . . . and many of these Jupiter-sized planets, or 'hot Jupiters,' *are* found orbiting closely, with some found within the habitable zones of their parent stars. It therefore follows that there is a possibility that many of these so-called 'hot Jupiters' may be a form of *sub*-brown dwarf in a close-in binary relationship with their host stars.

We claim, therefore, that Jupiter's place in our postulated Antique Solar System, sans the six Saturnian planets, would have seen it assume an orbit more in keeping with what is currently being observed in the cases of the so-called 'hot Jupiters,' as well as some sub-brown dwarfs currently seen circling other sun-like stars. This then presents us with a question: Was Jupiter once the original 'hot Jupiter' or *sub*-brown dwarf of our proposed Antique Solar System?

Other 'Jupiters' that show how our Jupiter Might Once Have Been

Before attempting a full assembly of our proposed Antique Solar System, there are two other extra-solar systems with Jupiter-like planets positioned in the habitable zone that bear mentioning. One is the exoplanet *HD 28185 b*, which is about 138 light-years away in the constellation of Eridanus, and the other is *HD 240210 b*, a hugely massive planet circling a star in the constellation of Cassiopeia. Both have calculated minimum masses that could elevate them at a later date to *sub*-brown dwarf status — both are definitely gas-giant planets at a minimum.

Below is a composite diagram adapted from data provided by the *Visual Exoplanet Catalogue* that plots just how ideally placed these two exoplanets would be within our own Solar System's so-called habitable zone.

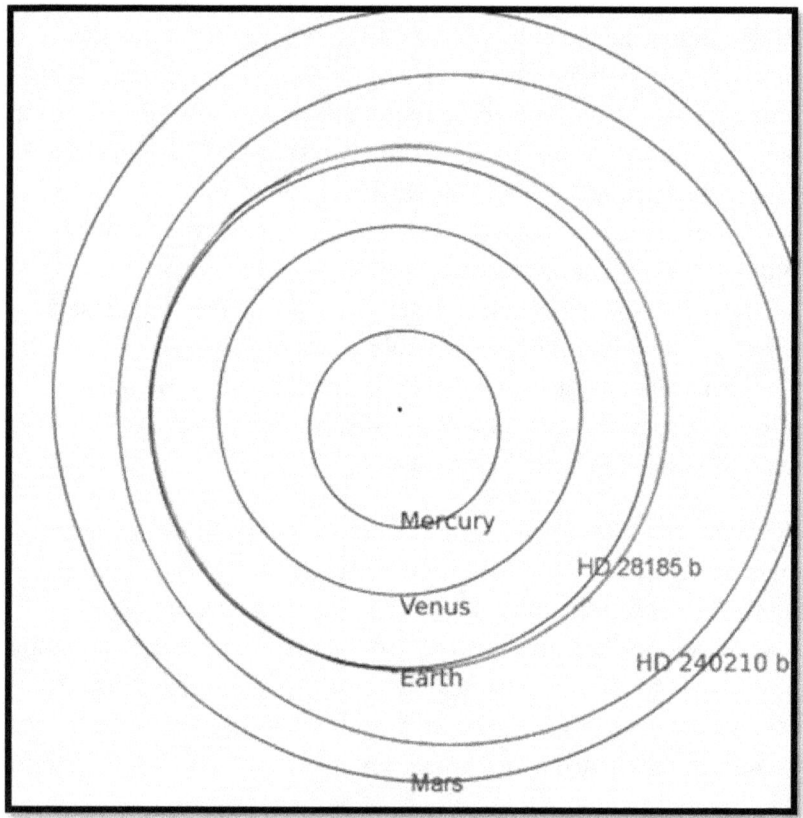

Both exoplanets HD 28185 b and HD 240210 b would orbit neatly within our Solar System's habitable zone, which is roughly between Earth and just outside Mars' orbit.

What this tells us, conclusively, is that stars like our own sun can, and most definitely do, support Jupiter-sized planets at orbits within the desired habitable zones. What we also know is that planets of these sizes are more than likely to support large terrestrial-type moons — remember that Jupiter has four large moons of its own. Is it possible then that exoplanets like *HD 28185 b* and *HD 240210b*, and suspected *sub*-brown dwarfs like *HD 141937 b*, may be the perfect harbors for extraterrestrial life on any moons they might support?

Is it then also possible that our own huge gas-giant, the massive Jupiter, once enjoyed an orbit in our sun's habitable zone and might have also been a safe haven for life on at least one of its four largest terrestrial satellites: the so-called famous Galilean moons of Jupiter?

Basking in the Warmth of a Sub-Brown Dwarf Star

Had Jupiter once been large enough to have once qualified as a brown dwarf, then its surface temperature could have been as high as 2,000 Kelvin. While only putting out a mere 112 – 165 Kelvin today, as a *sub*-brown dwarf, Jupiter's surface temperature would have probably been at the low end of brown dwarf stars; about 750 – 1,000 Kelvin. In its current orbit this would have been warm enough to increase temperatures on its two innermost terrestrial Galilean moons, the icy Europa and the volcanic Io. The two outer Galilean moons, Ganymede and Callisto, would still be fairly inhospitable places.

However, what this does not take into account is the EU understanding of a *sub*-brown dwarf as an electrically charged body moving through an electrical field. Had Jupiter been a *sub*-brown dwarf then it would be entirely expected to have developed its own cocooning plasma sheath, a kind of protective electric bubble that encapsulates all charged celestial bodies moving through electrical fields in space.[24] In fact, Jupiter has an enormous active plasma

---

[24] In the field of plasma cosmology a plasma sheath is otherwise known as a Langmuir sheath. This is the double layer (DL) sheath that Irving Langmuir discovered was created by plasma as a way of isolating electrically one section of itself from another. Charged celestial bodies moving through the vast electrically charged realms of space will form a

sheath even today, as was seen when comet Shoemaker-Levy 9 slammed into the planet back in 1992.

> "On 7 July 1992, SL-9 grazed past the giant planet Jupiter a mere 20,000 km above the cloud tops. It had penetrated deep into Jupiter's *huge plasma sheath*. As it switched suddenly from the Sun's electrical environment to that of Jupiter, it would have experienced extraordinary internal electrical stress. Unsurprisingly, it broke up."[25] (Emphasis ours)

Satellites inside this plasma sheath could be expected to have received an intensified boost from Jupiter's warming effects had it once been a *sub*-brown dwarf. The plasma sheath would have assisted by reflecting back much of Jupiter's own stellar energy, acting like a kind of blanket. It's then possible that the outer Galilean moons would have been warm enough for abundant liquid water. Otherwise, as is theorized by observers today, liquid water may have only existed on these moons as a subterranean ocean beneath deep icy crusts.

One thing we can say about this protective plasma bubble-like sheath surrounding our proto-*sub*-brown dwarf Jupiter is that it would not have been opaque, but as transparent as it is today, mostly due to the Sun's overwhelmingly positively charged presence. While other *sub*-brown dwarfs floating freely in space tend to display opaque plasma sheaths, Jupiter's own sheath would have been electrically equalized to the Sun's electrical field, a feature of it being inside the Sun's own giant plasma sheath, the heliosphere. This would have been the case no matter how far out its orbit was. As long as Jupiter remains inside the Sun's plasma sheath,[26] its own plasma sheath will remain transparent.

DL, or plasma sheath, which acts like a kind of cocooning bubble against foreign intruders. It should not be confused with magnetospheres.

[25] Wallace Thornhill and David Talbott, *The Electric Universe*, Mikamar Publishing, page 108.

[26] The Sun's plasma sheath is often referenced as its 'heliosphere'. A more accurate reference would be to what is called the Sun's 'heliopause'.

Nonetheless, as a *sub*-brown dwarf, Jupiter would have radiated much more intense warmth within its plasma sheath than anything we see emanating from the planet today.

Also worth taking note is that the actual appearance of Jupiter's warm *sub*-brown dwarf glow to human visual perception would be as deep coal tar magenta. The below comparison of the different classes of brown dwarf as seen by the human eye will help illustrate this more clearly:

Left to right - Sun, M-class brown dwarf, L-class brown dwarf, T-class brown dwarf, Jupiter. This is how these objects would appear to the human eye. Jupiter as a sub-brown dwarf would probably look more like the T-class object, which glows a dim magenta mostly due to its lack of light in the green portion of the spectrum due to absorptions by sodium and potassium atoms. **ARTWORK CREDIT**: *Dr. Robert Hurt of the Infrared Processing and Analysis Center*

If we then apply the observed orbital models of extra-solar Jupiter-like planets and place Jupiter in the Solar System's habitable zone, all kinds of interesting possibilities begin to present themselves. The most obvious of these is that Jupiter's three major ice-covered moons, Europa, Ganymede and Callisto, would see a radical change to their physical environments — liquid water, not ice, would become the dominant feature defining their surfaces (as long as there was a thick enough atmosphere). These former barren-looking satellites would undoubtedly be transformed into brightly lit, saturated water worlds fully capable of supporting aquatic life.

In fact, life on Jupiter's moons, ensconced as it were in their host *sub*-brown dwarf's protective plasma sheath and luxuriating in the added warmth of the Sun's radiant splendor, would find itself basking in an environment not unlike the fabled fictional world of Tatooine,[27] complete with its famous twin suns — except things would be much, much wetter.

Fog lifting on a two-sun daybreak. Artist's impression of Jupiter as a T-class sub-brown dwarf looking east from on high latitude on Ganymede's orbit-facing hemisphere. Ganymede is phase-locked to Jupiter like the moon is to Earth, so the same side is always facing Jupiter. Europa can be seen transiting Jupiter just above the horizon while Io is a small black dot almost lost in the corona of a much closer Sun. Small pumice rafts/bergs with vegetation dot the

[27] Tatooine is the fictional home planet for the characters Luke Skywalker and Anakin Skywalker of *Star Wars* fame.

The Radiation Issue

However, there would still be the problem of Jupiter's intense radiation, a feature of the great planet that is still with us today and can be relied upon to have been just as intense in the past had Jupiter once been a sub-brown dwarf. The Sun's own intense solar radiation, particularly its frequent flared outbursts, has long been a source of anguish for would-be inter-planetary space travelers looking to spend time out and away from Earth's protective magnetosphere.

Protected as we are on Earth by our planet's substantial magnetosphere, exposure to the Sun's solar radiation in deep space can prove fatal, especially if caught in one of the Sun's C-class or X-class flares. The same thing applies when travelling through Jupiter's own electrical environment, especially if it were functioning as a *sub*-brown dwarf. Brown dwarfs of all sizes are fully capable of flaring with the same deadly effects seen emanating out of our sun. Satellites in orbit around a close-in Jupiter would be subject to threats of intense flares spewing deadly radiation out into space from both the Sun and Jupiter.

A good protection against solar radiation, whether from the sun or Jupiter, is to envelope yourself in water. It has been a long-standing, if impractical solution for space travelers to seek ways to surround themselves with their craft's water tanks as a shield against solar radiation outbursts and flares. Aquatic life forms on Jupiter's three water saturated moons, Europa, Ganymede and Callisto, could theoretically enjoy this kind of protection from Jupiter's intense radiation. A decently thick atmosphere would also help, but more on that in another chapter. However, the best defense, and possibly the only reliable defense, is to have your own intrinsic magnetosphere. Of the water world moons of Jupiter, only one has the required intrinsic magnetosphere to ensure the safety of life on its surface . . . the ice-entombed moon called Ganymede.

In the next couple of chapters we will take a closer look at Jupiter's largest moon and assess its capacity to support life in a postulated Antique Solar System where Jupiter and its satellites orbit the Sun within the habitable zone. In doing so we will argue a case for Ganymede as not only the most likely place in the Antique Solar System for the emergence of life, but possibly the most likely place for the emergence of human life!

Summary and Takeaways from this chapter

Part II of this work is launched with the hypothetical question of how an Antique Solar System might have looked before the postulated arrival of a captured sub-brown dwarf star called Saturn and its accompanying satellites. (The subject of Saturn's arrival and its relationship to Earth is discussed at length in Part I of this work in the chapters "Discovering the Antique Solar System" and "Mankind's Purple Dawn").

- According to our scenario in answer to the above question, ancient Jupiter is elevated to the sub-stellar status of having once been a T-class sub-brown dwarf, enjoying a much tighter orbit around the sun within the latter's so-called 'habitable zone'. This revised orbital relationship is arrived at after assessing recent exoplanet data that shows growing numbers of Jupiter-like planets and suspected brown dwarfs in similar orbits around other sun-like stars. The data, we suggest, points to close orbiting Jupiter-like planets being the norm throughout the universe; this fact supports our hypothesis while contradicting currently accepted notions for gas-giant planet formation according to the accretion dictates of the Solar Nebular Disk Model (SNDM).
- The postulated ancient orbit of Jupiter within the Sun's habitable zone forces us to focus our attention on the effects this would have had on the four terrestrial Jovian moons referred to as the Galilean Moons. Three of these

182

four satellites are currently designated as ice moons, a condition that would have dramatically changed them to that of being liquid water moons had any of these Jovian satellites supported an atmosphere while that close to the Sun. This immediately suggests at least three ancient celestial objects that might have been capable of sustaining life as we know it. Importantly for our unfolding hypothesis, it should also be noted that these moons would have also been *bright* terrestrial worlds due to their close proximity to the Sun.

- Jupiter's reclassification as a former T-class sub-brown dwarf also provides a solution as to where all the water currently locked up as ice on the Galilean moons came from. Brown dwarf stars are known carriers of water and the misting and shedding of this water within a brown dwarf's plasma sphere can be expected to settle on any rocky satellites orbiting within. It is precisely this scenario that we suggest for the origins of the vast amounts of water ice found on the Jovian satellites.

- Finally, any suggestion of Jupiter's moons having once being capable of supporting life needs to address the issue of the substantial and destructive Jovian radiation that permeates the space taken up by the current orbits of the Galilean moons. One moon in particular, Ganymede, surprisingly provides the ideal defense against the deadly Jovian radiation by deploying its own intrinsic magnetosphere, the only known moon in the entire solar system to do so. Had Ganymede ever boasted an atmosphere too, then the suggestion is that Ganymede would have been the perfect place in our postulated Antique Solar System to sustain life as we know it — thoughts that are further developed in the next chapter.

Ganymede: Third Rock from Jupiter

We have been to see them up close, yet we still know very little about the moons that circle the Solar System's largest planet. Unmanned space probes have regularly made Jupiter and its moons a stopping-off point on their travels, the most recent one being the highly successful *Galileo* probe, which returned some of the best pictures to date of these enigmatic satellites. These probes have helped confirm that three of Jupiter's major moons are covered in thick icy crusts that hold out the potential for microbial life. If this proves to be the case, then the existence of life beyond the Earth would become a fact, something that would herald in a new understanding of our position in this vast universe.

The Galilean moons. From left to right, in descending order of size: Ganymede, Callisto, Io, Europa. Image Credit: NASA.

Back to the Jovian Moons

NASA's 2003 conceptual study dubbed Human Outer Planets Exploration (HOPE) identified the Jovian moon Callisto as the best place in our Solar System to build a habitable base aimed at facilitating further outer exploration of the Solar System.[1] It was

---

[1] See: Patrick A. Troutman (NASA Langley Research Center) et al., "Revolutionary Concepts for Human Outer Planet Exploration (HOPE)".

chosen largely on the basis that Callisto was well outside Jupiter's dangerously powerful radiation belt. On this moon, it was reasoned, human colonists would not be subject to the lethal effects of radiation poisoning suffered by the other Jovian moons. This scenario is, of course, predicated on the assumption that humans will have figured out how to deal with the high solar radiation levels that would be experienced by interplanetary travellers on their way to Callisto. And, in any case, Callisto is still subject to solar radiation bursts, as is the Earth's moon, due to its lack of an intrinsic magnetosphere. The study concluded it would probably be as late as the 2140s before such a project could become a reality.

Of the four so-called Galilean moons,[2] the innermost two, Io and Europa, are massively affected by Jupiter's intense radiation output. This is especially true of Io, which receives up to 3,600 rems[3] per day, hugely more than the 75 rems needed to induce radiation poisoning in a human.

---

[2] Galileo Galilei discovered Jupiter's four major satellites on January 7, 1610, and since then they have been referred to as Jupiter's Galilean moons.

   The naming of the moons was a fraught affair, with Galileo intent on first naming them after the children of his patron, the Florentine banker de Medici. However, Simon Marius, who had discovered the four moons independently at the same time, prevailed in having the satellites named according to a suggestion by Kepler after the lovers of Zeus; Io, Europa, Ganymede and Callisto. In a fit of pique Galileo refused to use these names and instead developed the numbering system in which he designated each moon from the closest outwards I, II, III, IV. This system was used right up until the 20th century until the discovery of lesser moons around Jupiter, at which point Marius' names came into general use.

[3] A rem is a unit measure of the radiation dose called a Sievert (Sv). 1 Sv = 100 rem. Exposure of to about 75 rems for a few days can cause radiation poisoning, while 500 rems over a few days is fatal.

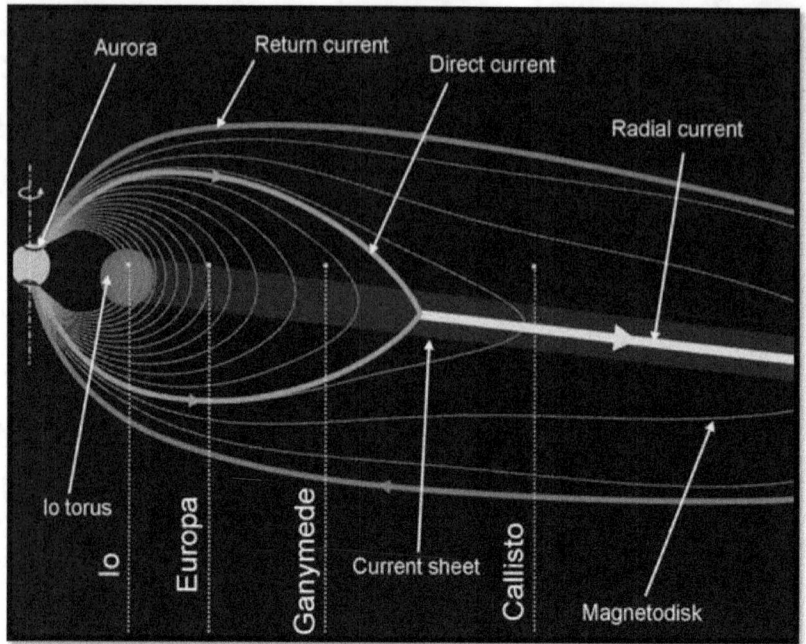

The magnetic field of Jupiter and co-rotation enforcing currents. Callisto's inclined orbit takes it outside the main current sheet of radiation, as does Ganymede's orbit. The 'Io torus' is a particularly heavy area of radiation within Jupiter's magnetosphere making exploration and colonization of Io extremely challenging. Image credit: RuslikO at Wikipedia.

Io is believed to be volcanically active, so much so that it has been claimed as the most volcanically active place in the Solar System.[4] This volcanism is ascribed to tidal heating, the effect of Io being pulled gravitationally by both Jupiter's gravity, and the gravity of the other Galilean moons. Under Electric Universe (EU) models, much of this volcanism can also be attributed to an intensely active electrical connection between Io and Jupiter, the plasma–rich Birkeland currents that exist between most of the planets and the

---

[4] E.g.,
http://www.planetaryexploration.net/jupiter/io/volcanism_on_io.html

Sun and between large gas-giants and their satellites.[5] In the meantime, Io orbits inside a particularly intense torus of radiation making it all but impossible for human colonization.

Europa is not that much better for human habitation, itself receiving up to 540 rems of radiation per day. But Europa has a thick crust of ice, and the presence of this shell of solid water has sparked all kinds of speculation on the possible existence of microbial life deep under the ice in a suspected hidden interior liquid ocean. Exciting as this may seem, the radiation issue, along with an extremely low gravity, still poses a significant barrier to the establishment of a human colony on Europa.

However, Europa's problems didn't stop a group called the Artemis Project from proposing a plan in 1997 for the colonisation of Europa.[6] To combat the severely cold conditions on Europa's surface, the Artemis Project proposed that the colonists drill down to under the icy curst to establish a base nearer to where liquid water might be. Here they would be shielded by the ice from the radiation, yet possibly remain warm enough to survive the frigid environment. Supposedly they could then set about settling in with any microbial neighbours for the long term, especially since there appears to be sufficient oxygen on Europa to support life.

Fascinatingly, in his science fiction novels *2010: Odyssey Two* (1982) and *2061: Odyssey Three* (1988), Arthur C. Clarke explores the idea of an intelligent alien species kick-starting Jupiter's transformation into a star, which then transforms Europa into a tropical ocean world from which humans are banned (the story seems to reflect a particular bias within the space exploration establishment towards the investigation of Europa when it comes to the Jovian moons, a theme that has been reflected in science fiction for some time.)

---

[5] http://www.planetaryexploration.net/jupiter/io/io_plasma_torus.html
[6] Kokh, Peter; Kaehny, Mark; Armstrong, Doug; Burnside, Ken (November 1997). "Europa II Workshop Report". *Moon Miner's Manifesto* (110).

Skipping past Ganymede to Callisto for the moment, Jupiter's outermost Galilean moon does indeed appear an enticing destination for human colonisation — if it were not for the fact that it is so incredibly cold. Safely out of Jupiter's troublesome radiation zone, Callisto is another potential water world where microbial life may exist. However, this moon's meagre atmosphere is mostly made up of carbon dioxide, not the best mix for human habitation. While microbial life and flora would flourish in such an atmosphere, the lack of oxygen would be problematic for long-term human colonization.

The potential for microbial life on Callisto seems slim, mostly due to the extremely low temperatures currently experienced on its surface. According to accepted mainstream theory, Callisto's surface belies its ancientness with supposed impact craters testifying to a long history of periodic and intense cosmic bombardments.

> "The ancient surface of Callisto is one of the most heavily cratered in the Solar System. In fact, the crater density is close to saturation: any new crater will tend to erase an older one. The large-scale geology is relatively simple; there are no large Callistoan mountains, volcanoes or other endogenic tectonic features."[7]

With Jupiter believed to be acting as a giant gravitational vacuum cleaner for the Solar System, the chances of future bombardments remain high, especially since Callisto is so far outside Jupiter's massive magnetosphere. This would point to a tenuous existence for any human colony established on Callisto's surface.

---

[7] Wikipedia entry for "Callisto", see http://en.wikipedia.org/wiki/Callisto_(moon)#Surface_features, This entry cites; Zahnle, K.; Dones, L. (1998). "Cratering Rates on the Galilean Satellites" (PDF). *Icarus* **136** (2): 202–222. Bibcode 1998Icar..136..202Z. doi:10.1006/icar.1998.6015. PMID 11878353. http://lasp.colorado.edu/icymoons/europaclass/Zahnle_etal_1998.pdf; and Bender, K. C.; Rice, J. W.; Wilhelms, D. E.; Greeley, R. (1997). *Geological map of Callisto*. U.S. Geological Survey. http://astrogeology.usgs.gov/Projects/PlanetaryMapping/DIGGEOL/galsats /callisto/jcglobal.htm.

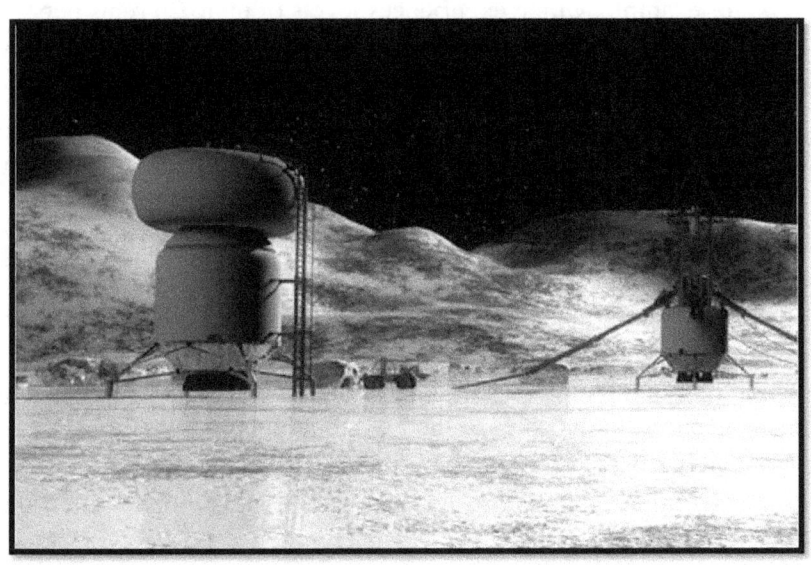

*Artist's impression of a future human base established on Callisto.
Here they could supposedly mine for the fuel needed to supply
spacecraft being sent on to the furthest reaches of the Solar System.
Credit: NASA*

This brings us back to Ganymede, the largest of the four Galilean moons, and indeed the largest moon in the Solar System. Bigger than the planet Mercury and at about three quarters the size of Mars, Ganymede would provide the best gravitational environment for humans of the Jovian satellites. It enjoys a meager oxygen-rich atmosphere and is absolutely saturated in water, which currently forms a global ice crust encapsulating its entire surface.

As noted in the previous chapter, Ganymede also enjoys the protection of its own intrinsic magnetosphere, the only moon in the Solar System to have one. Despite being inside Jupiter's substantial and highly dangerous magnetosphere, Ganymede's native magnetosphere significantly cuts down the Jovian radiation to an entirely manageable 8 rems per day.[8] This is still high compared to Mars, for example, which experiences levels of up to 20 rems per year (.06 rems per day). However, even without the protective properties of an atmosphere, humans with adequately shielded apparel and vehicles could travel about on Ganymede's icy surface with little concern for the long-term effects of Jovian radiation.[9]

The cause of Ganymede's magnetosphere is still a mystery to mainstream planetary scientists, who continue to look to variations on the internal 'dynamo' theory where convection within a planet or moon's inner core area is posited as the source of the electrical energy needed to establish a permanent and intrinsic magnetosphere. Advocates of the EU model merely point to Birkeland currents as the source of this electrical energy in providing magnetism in all its forms, wherever it is found in the Solar System and beyond.[10]

---

[8] Frederick A. Ringwald (2000). "SPS 1020 (Introduction to Space Sciences)". California State University, Fresno.

[9] NASA limits its astronaut's direct exposure to cosmic radiation to 600 rem for their career, a maximum of 300 rem per year, and a maximum of 150 rem per 30 days. A year on Ganymede would amount to exposure in the realm of approx.. 2,900 rem, a total well beyond NASA's current safety limits. The 1991 SIMM class decided on a limit of approx. 50 rem/year, which matches the US' National Council on Radiation Protection and Measurements (NCRP) recommendations. Obviously, in the absence of a protective atmosphere, human habitation of Ganymede would still require advanced radiation shielding to be viable over the long term. For details on NASA radiation limits for astronauts see: http://people.ee.duke.edu/~chappell/mars/radhuman.html

[10] Norwegian scientist Kristian Birkeland discovered the concept of large electrical currents flowing into the Earth from space over a hundred years ago. In his famous terrella experiments, Birkeland was able to demonstrate at the laboratory level how aurora effects are created by directing electrical currents through metallic balls suspended in a vacuum. The sheer inability of scientists of his day to perceive where this electrical current could come from severely restricted the impact of Birkeland's ground-breaking experiments in suggesting an outside source for the Earth's magnetic field. Plasma physicists such as Hannes Alfvén have subsequently identified Birkeland currents as the source of the necessary electrical currents needed to establish magnetic fields. A common refrain of the Electric Universe

Ganymede's intrinsic magnetosphere is engulfed in Jupiter's massively larger and more powerful magnetosphere, yet manages to cut Jovian radiation levels down to 8 rems on the satellite's surface. The Electric Universe model points to large and electrically active Birkeland currents flowing between the satellite and Jupiter as the true source of Ganymede's magnetosphere. Credit: NASA/European Space Agency.

Interestingly, of the four ice-covered Galilean moons, Ganymede has received the least amount of interest as a possible future human colony for continued outer solar system exploration. Even layman-type wiki websites aimed at the general public and devoted to

---

model is that electrical current is always needed to initially create magnetism and that the copious magnetic fields known to exist in space are merely the ashes of vast electrical currents that have once flowed, or continue to flow in space.
See: Wallace Thornhill, "Alfvén Triumphs Again (& Again)", May 9, 2011, http://www.holoscience.com/wp/alfven-triumphs-again-again/

discussing various scenarios for the colonization of space have little to say on the subject.[11]

A possible reason for this may be the perception that Ganymede is what is called a 'highly differentiated body'. Being a differentiated body means that Ganymede's composition is believed to be made up of distinctly separate layers; an iron core at its center surrounded by a rocky silicate core, and then a huge watery/icy mantel layer that is encapsulated by an extremely thick crust of ancient ice. While such a large layer of water may hold out hope for microbial life, the appeal of the moon as a future base for essentially land-based creatures like humans seems uncomfortably limited.

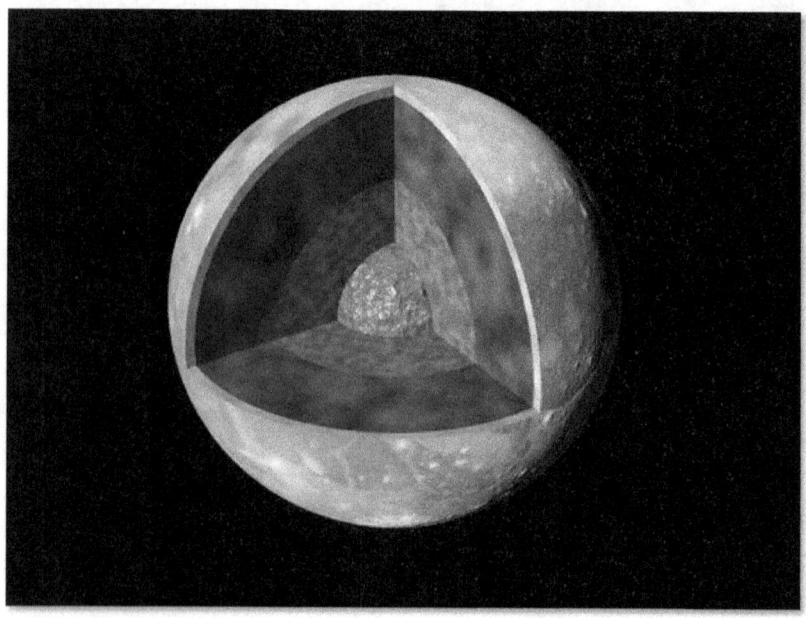

The mainstream consensus view of Ganymede's structural composition. An iron core makes up the central part of the moon and is wrapped in a silicate outer core. The next layer (blue layer) is a hugely deep icy mantel with a possible deep water ocean trapped between it and a thick icy crust, that forms Ganymede's

[11] See for example:
http://spacecolonization.wikia.com/wiki/Colonization_of_Ganymede

Another reason for this reticence in considering Ganymede as a future destination for human colonization is that Ganymede's differentiated composition contrasts it sharply with Callisto and Europa, both of which have been considered candidates for mining operations. Ganymede's supposedly 800 – 1000 kilometre thick ice mantel seems to offer little in the way of mining potential, especially in accessing the important types of minerals and fuels needed by projected deep space missions. Furthermore, speculations of a liquid ocean between Ganymede's icy crust and its ice mantel give this tentative body of water depths of up to 200 kilometres[12] — an impossible depth for known mining technologies to plumb in the search of materials essential for the sustaining of life on such a forbidding body.

The origin of this perception of Ganymede as a highly differentiated body comes from the data sent back by space probes making fly-bys of the Jovian satellite. In particular, the *Galileo* craft seemed to confirm that Ganymede had an almost impossibly low moment of inertia, making it the most centrally condensed body in the Solar System.[13] It is this low 'moment of inertia' that has led scientists to develop the compositional model for Ganymede that has it internally separated into such distinctly different layers. However, as we shall see, there may be an alternative interpretation as to why Ganymede has this extremely low moment of inertia, one that still points to Ganymede as a water world, but not with the supposed extreme depths of ice and oceans as proposed by the mainstream model.

---

[12] Solomonidou, A., Coustenis, A., Bampasidis, G., Kyriakopoulos, K., Moussas, X., Bratsolis, E., Hirtzig, M., "Water Oceans of Europa and Other Moons : Implications For Life in Other Solar Systems," *Journal of Cosmology, 2011, Vol 13. 4191-4211,* see: Introduction, paragraph 4.

[13] http://en.wikipedia.org/wiki/Ganymede_(moon)

The Enigma of Ganymede's Moment of Inertia

"Ganymede is about the lowest I/MR^2 [i.e., moment of inertia] conceivable for any solid object in the Solar System."[14]

The best way for the layman to understand what is meant in physics by the term 'moment of inertia' is to visualise an ice skater spinning on the spot — the more they pull their arms in the easier it is for them to spin faster. What they are doing is centralising their mass around the axis of their spin, or, in other words, condensing their mass inwards to pack it as tightly and densely as they can at the centre of their spin. By doing this they don't *have* to spin faster, but it certainly makes it easier to spin faster if they want to. Ganymede is the space equivalent of an ice skater who has pulled in his/her arms while going into a spin — it would be much easier, energy-wise, to increase Ganymede's spin rate than that of any other body in the Solar System.

Put technically, a body's moment of inertia is a property of the distribution of its mass in space that measures its resistance to rotational acceleration about its axis. It was the fly-bys of the space probe *Galileo* that confirmed Ganymede's moment of inertia exists at an almost impossible 0.3105. By calibrating the gravitational effect on Galileo as it swooped past Ganymede, scientists were able to determine this exceptionally low number. In comparison, the similar sized moon Callisto enjoys a moment of inertia of about 0.359, much higher than Ganymede.[15]

This has led scientists to proclaim that Ganymede has most of its heaviest and densest matter or material packed into its inner core. As noted above, iron has been identified as being at the center of the core with a separate rock silicate forming an outer core or mantel,

---

[14] See online Caltech course: "Ge 131: Planetary Structure and Evolution -- Spring 2012", Textbook PDF, chapter 11, page 124. http://www.gps.caltech.edu/classes/ge131/

[15] Ibid. page 125.

the two forming the bulk of Ganymede's 'density.' Wrapped around this is whatever fills out the rest of the moon till it reaches its observed volume size. This is believed to be mostly water in the form of ice, a belief that has led to proclamations that an ocean may have been discovered under Ganymede's icy crust.[16]

Whatever this outer mantel is made of it has to be far less dense than the inner two cores in order to give Ganymede its low moment of inertia. Water is the low-density substance favored by accepted theories, mainly because, if the water is salty, it would possibly also serve as an adequate inductor contributing to Ganymede's observed intrinsic magnetosphere. According to this thinking a silicate-based rock-like outer mantel would be too dense and heavy to account for Ganymede's low moment of inertia while also making for a poor conductor for electrical induction.

> "Ganymede's internal structure appears to include a metallic core, a rocky mantle, and an icy outer layer, a model inferred from measurements of the gravitational moments (Anderson *et al.* 1996) and magnetic data (Schubert *et al.* 1996, McKinnon 1997). *An inductive response could be present if the icy layer contains electrically conducting paths as, for example, in regions of partial or complete melt of sufficient thickness.*"[17]
> (Emphasis ours)

The above quote brings up an important point, i.e., the electromagnetic conductivity of whatever Ganymede's supposed massively deep icy mantel is made of. Salt water is the prime candidate because of its known conductive properties. Add to this its compatible density, which conforms to Ganymede's low moment of inertia, and it seems obvious that it is indeed salty water that constitutes this outer mantel. But, as we shall soon see, there is

---

[16] M. G. Kivelson, K. K. Khurana and M. Volwerk, "The Permanent and Inductive Magnetic Moments of Ganymede", PDF, Icarus 157, 507–522 (2002), doi:10.1006/icar.2002.6834

[17] Ibid, page 2.

another compound that can substitute for salt water, a solid rock-like compound just as potentially conductive and, actually, less dense.

But first, it should also be noted that the presence of a huge mass of water also seems to conform to accepted concepts of how planets and moons in general supposedly evolve. But the seemingly absolute separation of each of Ganymede's layers contrastingly and actually poses a real problem to this model. Under the accepted thermal cooling principles of planetary formation, the outer layers of any given body in the Solar System should cool quicker than the inner layers. However, Ganymede's highly differentiated layers implies that its cooling period has not behaved in the way most thermally-based evolution models would suggest, a process that would normally produce much more mixed layers than is believed to exist on Ganymede. In summary then, Ganymede's low moment of inertia seems to pose more problems to the accepted model of planetary and moon formation than most would probably care to admit. For example, Michael Thomas Bland has noted in a dissertation on the subject:

> "Jupiter's satellite Ganymede is one of the Solar System's great enigmas. Gravity data indicate that it is the most centrally condensed solid object in the Solar System, suggesting a high degree of differentiation (Anderson et al., 1996). Additionally, the satellite is one of only three solid bodies in the Solar System that produces an internally driven magnetic field. Finally, its surface bears the scars of vigorous tectonic activity that were likely created in the middle of the satellite's geologic history. Furthermore, all these features occur on a satellite that currently receives no tidal heat and whose "twin" (in size and bulk composition) Callisto is only partially differentiated (Anderson et al. 1997b) and dominated by impact cratering. *Scenarios for Ganymede's thermal evolution that assume secular cooling of the satellite over the age of the Solar System*

have difficulty explaining the observations described above (e.g. Freeman, 2006)."[18] (Emphasis ours)

When it comes to the assumptions and methods employed by mainstream science in determining the composition of other bodies in the Solar System, EU physicist Wallace Thornhill has this extra caveat from an EU perspective to add to the discussion about the masses involved in making up celestial objects:

> "We conceal our ignorance of any underlying physical mechanism by tolerating dimensional constants. If mass is an electrical variable, G[gravity] cannot be constant. Assuming G to be universal as well gives rise to calculated masses and densities of celestial bodies that lead to further conjectures cantilevered upon the already dubious assumptions. _Stellar and planetary structure and composition are based upon this erroneous conviction._ For example, by using G, measured on Earth, the planet Saturn appears to have a lower density than water!"[19] (Emphasis and underlining ours)

There is something obviously different in the way Ganymede was formed when it is compared to other moons and planets in the Solar System. So, the question must be asked: is there any other possible internal composition that could explain Ganymede's highly differentiated layers, yet extremely condensed centre? Is there yet another substance that can account for Ganymede's seemingly low-density outer mantel that explains its low moment of inertia?

We believe the answer lies in the formation of a solid silicate-based substance that is formed at high temperatures, yet can assume a

---

[18] Michael Thomas Bland, "The Tectonic, Thermal and Magnetic Evolution of Icy Satellites", *Phd dissertation submitted to the Faculty of the Department of Planetary Sciences, The University of Arizona*, page 20. See Google books: http://books.google.co.uk/books?id=--FdM9ep6coC&printsec=frontcover&source=gbs_ge_summary_r&cad=0#v=onepage&q&f=false

[19] Wallace Thornhill, "Newton's Electric Clockwork Solar System," April 21, 2009.

density far less than water. It is a rock, and this rock's name is *pumice*.

A Ball of Pumice Wrapped Around an Iron Core

The suggestion that Ganymede's outer mantel may consist of pumice with an icy crust sitting on top of it would elicit a number of obvious objections from mainstream scientists, the least being that pumice is generally a low-density rocky product born of volcanic activity. It is therefore highly improbable that an 800 – 1,000 kilometre layer of pumice could be formed purely by volcanic eruptions. The other Jovian moon Io, as noted, is the recognised champion of volcanism, and we do not see an 800 kilometre layer of pumice on that moon. Unfortunately, Ganymede's outer mantel needs to approach these 800 kilometre depths to account for the satellite's low moment of inertia — so what gives with the idea that it's made of pumice?

On Earth, ". . . pumice is created when super-heated, highly pressurized rock is violently ejected from a volcano. The unusual foamy configuration of pumice happens because of simultaneous rapid cooling and rapid depressurization. The depressurization creates bubbles by lowering the solubility of gases (including water and CO_2) that are dissolved in the lava, causing the gases to rapidly exsolve (like the bubbles of CO_2 that appear when a carbonated drink is opened). The simultaneous cooling and depressurization freezes the bubbles in the matrix."[20]

It's these bubbles frozen into the rock that makes pumice so light and buoyant in water. Immense thin rafts of pumice with area sizes larger than the state of Israel have formed in the Pacific after major eruptions.

> "Pumice is composed of highly microvesicular glass pyroclastic with very thin, translucent bubble walls of extrusive igneous rock. It is commonly, but not exclusively of silicic or felsic to intermediate in composition, . . . but

---

[20] Wikipedia entry on "Pumice," See: http://en.wikipedia.org/wiki/Pumice

basaltic and other compositions are known. . . . It forms when volcanic gases exsolving from viscous magma nucleate bubbles which cannot readily decouple from the viscous magma prior to chilling to glass."[21]

From the above quote then, we know that heat, and not just any heat, but super heat is needed in the formation of pumice and that pumice can be produced from a number of different super-heated rock types. Therefore, if a planetary body the size of Ganymede was originally made up of an iron core wrapped in a larger silicate rocky mantle, then we would need a huge portion of that rocky mantle to be super-heated to a great depth on a global scale to produce an outer layer of pumice. Clearly, internal volcanism within Ganymede could not account for such a condition. The question then remains if there is a mechanism where a whole moon's outer mantle can be super-heated to create an 800 kilometre layer of pumice?

Plasma in arc mode can produce the kind of heat we are talking about. The Sun, a star engulfed in plasma crackling in arc mode, regularly ejects flares and plasma filaments with temperatures far in excess of that needed to melt rock. This, then, opens up a number of possibilities when we factor in the idea of Jupiter having once been an electrically active sub-brown dwarf, as discussed in the previous chapter. Could Jupiter have once emitted an enormous and sustained super-heated electrical flare or electrical current that engulfed Ganymede in a global plasma fireball capable of rendering Ganymede's original outer rocky mantle into pumice?

The scenario suggested in the above question would call for Jupiter to have experienced an electrical overload capable of initiating an extreme discharge approaching that of a nova event. Such events, especially X-ray flares,[22] are entirely plausible in the lives of brown

---

[21] Ibid.

[22] See: "First X-ray Flare from Brown Dwarf Observed," *News Release, July 11, 2000*, Office of Public Affairs, UC Santa Barbara. Also, see: "Brown dwarfs do form like stars (12/6/2008)," *Astronomy Report*,

dwarfs, especially under the tenets of the EU model. In fact, the EU concept of planet birthing calls for exactly this type of scenario, and it would adequately account for not only Ganymede's unique internal characteristics, but also for the dichotomy seen in the different structures of Jupiter's four Galilean moons. But more on that shortly. . .

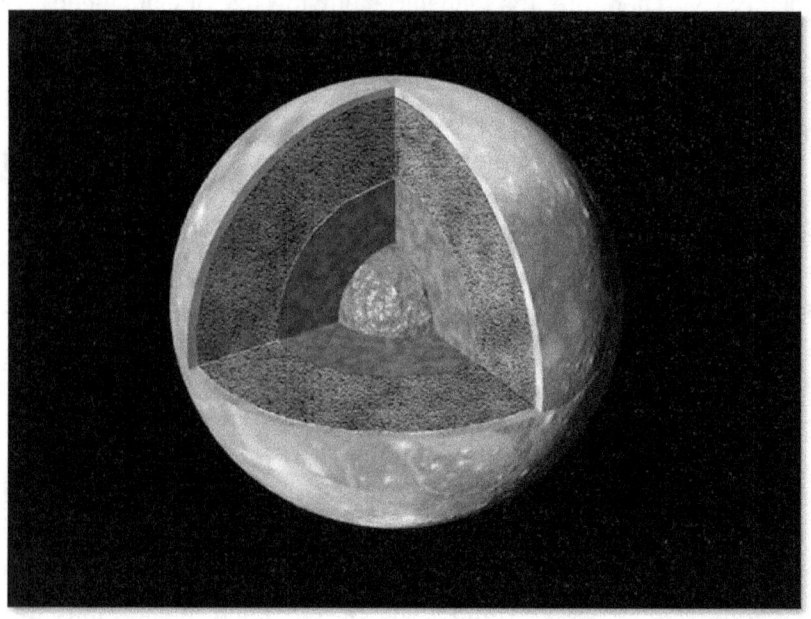

A revised model for the interior for Ganymede showing the outer mantel as a deep layer of low-density pumice encapsulating an inner core of iron and an outer core of silicate. Such a layer comprised of pumice would account for Ganymede's low moment of inertia just as easily as a deep icy ocean.

Ganymede's Electrical Conductivity: It's in the P-Holes

One final component worth mentioning, since the conductivity of salt water plays a major part in the mainstream identification of

http://www.astronomyreport.com/Research/Brown_dwarfs_do_form_like_stars.asp

Ganymede's outer mantel as salt water ice, is the growing acceptance in the geologic sciences of a concept called p-holes. P-holes are linked to the plasma-like phenomenon of earthquake lights and are said to be the means by which electricity can flow through rock to produce these mysterious lights.

> "The best theory of earthquake lights comes from mineral physicist Friedemann Freund, who took a frustrating fact [rocks samples conducting electricity along their surface in the lab] and made it the cornerstone of a new hypothesis. Under the conditions of most earthquake faults, lab experiments show that rocks are electrical insulators. But the experiments are very frustrating because rock samples conduct electricity well along their surfaces. This swamps the effects being looked for, unless the samples are first roasted in vacuum. But at that point you don't have realistic rocks.
>
> "Freund realized that mineral grains in ordinary rocks are naturally full of flaws; specifically, oxygen atoms in imperfectly ionized states. There are millions of oxygen atoms in every piece of silicate mineral with one electron short, bound together in peroxy bonds. When such a bond is broken for whatever reason, the result is a pair of "holes" of positive charge, or p-holes. Anyone who knows the physics of semiconductors knows about holes. They carry charge, in their way, just as effectively as electrons do."[23]

Pumice is full of holes thanks to its unique formation — that is what makes it lighter than water. Though there are no confirmed tests on pumice itself concerning the presence of p-holes in the substance, it is inconceivable that pumice would not also exhibit this p-hole effect during its formation and therefore act in a similar way to salt water in the conducting of electrical currents. If so, then pumice is as complementary to the formation and preservation of Ganymede's magnetosphere as salt water is thought to be.

---

[23] Andrew Alden, "Earthquake Lights," *About.com Guide*, first published 5 May, 2006. see: http://geology.about.com/od/earthquakes/a/EQlights.htm

An Electrifying Scenario: The Birth of Jupiter's Moons

Within the EU model the birthing of planets takes place when a star, defined as an electrical discharge phenomenon, is electrically overloaded by the galactic current, which then results in stress fractures to its internal heavy element iron core. This leaves two positively charged pieces of the core at the centre of the star, with the lesser piece being forcibly ejected/repelled by the like-charged (magnetised) larger piece. The smaller piece of iron core then finds itself outside its star's interior where it assumes an orbit. It is now a planet and the dusty debris of the star's circumstellar disk will soon be attracted to the smaller piece's existing magnetic field, which begins the process of forming layers. That is how mantels are constructed according to this theory.

Had Jupiter once been a sub-brown dwarf star, it can be expected that its four major moons were produced in this manner. While we are not saying that all four moons were ejected at the same time, it is conceivable that the first moon to find itself in orbit around Jupiter would have been severely affected by the birth of any subsequent moon, both gravitationally and electrically.

Accordingly we can now envisage the birth of, let's say Io and Europa, as having initiated huge super-heated Jovian jets and flares of plasma capable of literally incinerating an existing close-orbit moon, let's say Ganymede. Any Birkeland current connecting Ganymede to Jupiter during this proposed process would also go into electrical arc mode and contribute significantly to the super-heating of Ganymede's outer rock mantel. Then, in the instant that Jupiter's electrically charged environment reverted to glow mode as the nova event subsided, Ganymede's super-heated outer mantel would suffer simultaneous rapid cooling and rapid depressurization leading to the formation of an outer pumice mantel.

The above scenario would then see a former solid rock-like moon reduced to being an iron ball wrapped in a silicate mantel with an outer, deep layer of pumice. Large pieces of pumice at the surface would be fractured off the outer surface where they would lay like so much gigantic and light-weight rubble within the surface's residual dust. In time, water from Jupiter's sub-brown dwarf atmosphere would mist out into the confines of Jupiter's plasma sheath and ultimately mist down onto Ganymede. Eventually, as more water vapour collected and condensed on Ganymede's surface under a developing atmosphere, a liquid ocean would rise, and with it the buoyant pieces of fractured pumice littering the moon's landscape. Free-floating pumice islands of many different sizes would then drift across a global ocean forming giant archipelago-like rafts.

Concurrent with these developments, Ganymede would take on the role of cathode to Jupiter's electrical anode glow. Subsequently, electrolysis of the existing water vapour would separate Jupiter's H^2O molecules into hydrogen and oxygen. The hydrogen would be drawn toward the anode, Jupiter, while the oxygen would begin to collect around the cathode, Ganymede. Thus Ganymede's oxygen atmosphere would be born to provide the final component for the existence of liquid water on Ganymede. A climate would eventually develop on Ganymede and soon, possibly quickly, the existence of life under Jupiter's dull glowing mass and the Sun's more distant warming rays could be considered. . .

But that is for another chapter.

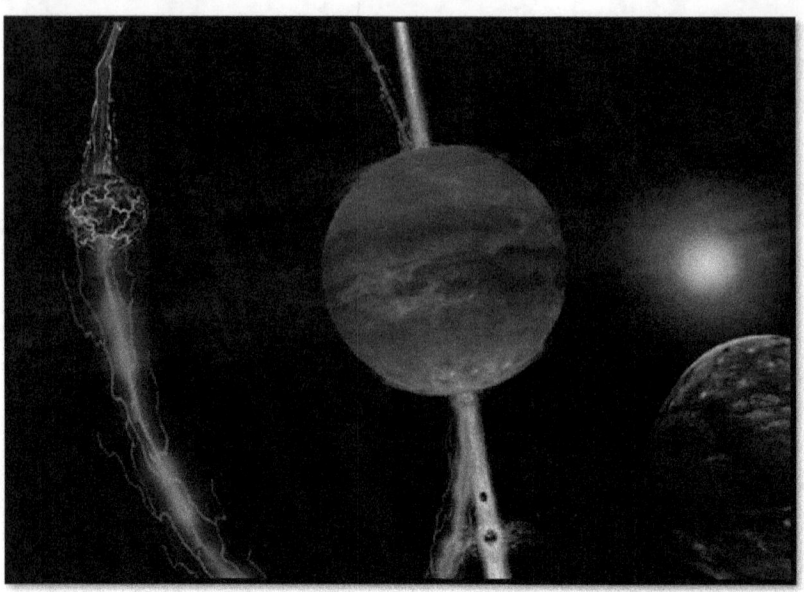

The birth of Io and Europa and the super-heated electrical baking of Ganymede.

In the distant past the sub-brown dwarf Jupiter (center) brightens while undergoing a nova-like flaring event, which produces violent polar jets that eject two solid bodies from Jupiter's interior – the new moons Io and Europa (dark dots seen in Jupiter's southern polar jet). Flaring plasma filaments super-heat Jupiter's circumstellar disk and engulf the orbiting Ganymede (at left of frame) — as does the intense electrically arcing Birkeland current connecting Ganymede to Jupiter, turning Ganymede into a super red-hot ball of molten rock being baked through. This may have gone on for weeks or months, and then stopped as suddenly as it started. With the subsidence of Jupiter's electrical flaring, Ganymede would have experienced a rapid cooling and depressurization leading to the formation of an outer mantel of pumice.

Callisto, a much older moon seen in the foreground at bottom-right of frame, is orbiting far enough out not to be affected as badly as Ganymede, but still experiences intense aurora effects and possible bombardments. The event is depicted as happening during the primordial Antique Solar System epoch when Jupiter is conjectured as having been in a much closer orbit to the Sun. The Sun, seen in the distant background, is about 155,000,000 kilometers away (96,300,000 miles).

Image not to scale.

Summary and Takeaways from this chapter

Jupiter's famous Galilean moons have been assessed in this chapter as places of interest within our solar system in the search for life beyond Earth. While mainstream science favours Europa for the potential existence of microbial life and Callisto for its potential as a future human outpost in the exploration of the outer solar system, we find Ganymede to be the true dark horse contender in the search for evidence of extra-terrestrial life. Ganymede's unique and intrinsic magnetosphere means this moon most closely resembles Earth in terms of the ability to provide a safe haven for life from the damaging effects of Jovian, solar and galactic cosmic radiation.

- While a mystery to mainstream science as to how Ganymede sustains its own magnetosphere, we point to the Electric Universe model and its recognition of current carrying Birkeland currents as the solution to how Ganymede powers its intrinsic magnetosphere. Ganymede's electrical relationship with an electrically active Jupiter is responsible for this unique feature, a feature that is analogous to Earth's own electrical relationship to the Sun in producing Earth's magnetosphere.

- Ganymede's almost impossible low 'moment of inertia' has lead mainstream planetary scientists to conclude that Ganymede is a highly differentiated body with a mantel that is believed to be almost exclusively made up of icy water to a depth of over 800 kilometres. Our suggestion is that Ganymede's mantel is actually comprised of lightweight pumice-like silicate created when Ganymede was electrically baked in the course of a flaring episode during Jupiter's sub-brown dwarf phase. A pumice mantel, being lighter than water, would more than adequately account for Ganymede's low moment of inertia.

- While mainstream consensus believes Ganymede's ability to maintain its magnetosphere is due to a conductive saltwater mantel, we suggest the phenomenon of electrically conductive 'P-holes', now being confirmed in rocks, as the more likely solution to Ganymede's mantel being conductive. Pumice, by its nature, is full of holes, and this feature makes it extremely susceptible to being a harbour for electrically conductive P-holes. A predominately pumice-like mantel on Ganymede provides a solution for both its conductivity in sustaining Ganymede's intrinsic magnetosphere while also accounting for its low moment of inertia.

- During the ancient formation of its pumice-like mantel by electrical discharges thrown out by an unpredictable and flaring Jupiter (then in close orbit with the Sun), Ganymede's surface would have come to be littered by the debris of broken shards of pumice and copious amounts of silicate dust. As water from Jupiter eventually condensed on Ganymede, along with oxygen molecules produced by electrolysis of Jupiter's water vapour, an atmosphere and ocean would have arisen on Ganymede's surface. The left over dust and broken shards of pumice would have risen with the rising global ocean to form free-floating island land masses.

What Happened to Ganymede?

There is a tendency in modern planetary science to believe that planets are where they are, and how they are, because that is where they were originally formed. The same reasoning is applied to planetary moons and satellites. But how reliable is this train of logic? And how can it be applied to what we see today on the surfaces of planetary bodies and their moons?

The process by which Ganymede's surface became a thick icy crust grooved and sputtered with seemingly innumerable impacts and segmented into distinct light and dark zones is assumed by mainstream consensus to have been the result of billions of years of exposure to roughly the same orbital environment that Ganymede currently now enjoys. Added to this is the assumption that the non-impact features seen on Ganymede's surface must surely be the result of some form of long-term internal tectonic activity affecting the satellite's thick icy crust. For example:

> "Ganymede has fascinating and diverse surface geology. Its relatively young bright "grooved terrain" is shaped by Earth-like tectonism, with faults and fractures that deform its surface ice. Icy volcanism may have paved its smoothest terrains. Its ancient cratered "dark terrain" shows a complex geological history. The dark terrain probably dates from the earliest days of the Galilean satellite system and offers the best promise for unravelling its cratering history."[1]

Currently, the internal forces and tidal forces needed for this tectonic activity seem to be subdued almost to the point of being non-existent. There is little evidence of this type of activity taking place on Ganymede. Planetary scientists have thus countered by

---

[1] R. T. Pappalardo, K. K. Khurana, and W. B. Moore, "THE GRANDEUR OF GANYMEDE: SUGGESTED GOALS FOR AN ORBITER MISSION," Brown University (Dept. Geological Sciences, Providence RI 02912-1846; 2UCLA (Los Angeles, CA 90095-1567), 4065.pdf, see: http://www.lpi.usra.edu/meetings/outerplanets2001/pdf/4065.pdf

suggesting that Ganymede must have once experienced periods of huge internal tidal effects caused by previous unstable orbital resonances produced over time by its evolving relationship to the other three Galilean moons. But can this really account for all the features seen on Ganymede?

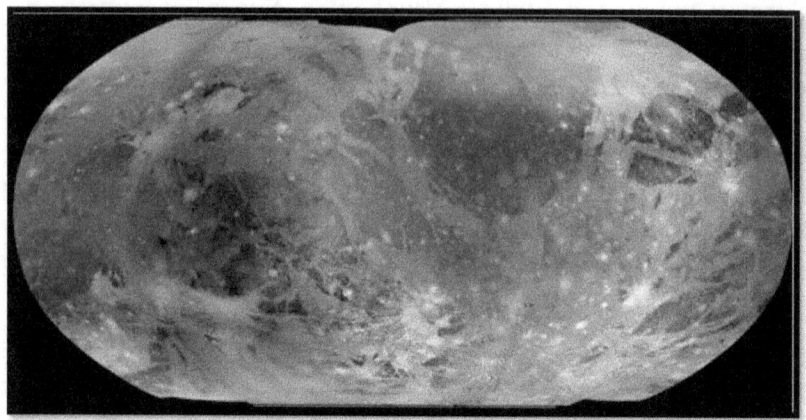

A flattened projection of Ganymede's surface. The darker areas are considered to be the Jovian satellite's more ancient terrain while the lighter areas represent newer resurfacing by supposed tectonic activity. The bright speckled spots are believed to be impact craters with a few larger supposed impact craters showing extensive radial patterns. Credit: NASA.

An interesting feature of the three innermost Galilean moons is that they share the so-called Laplace resonance of 1:2:4. That basically means that for every completed orbit of Ganymede, Europa will complete two orbits and Io will complete four. It is theorised that whatever may have brought these three moons into this resonance may have also have been responsible, via extreme tidal kneading, for creating the internal forces at Ganymede's core capable of producing the supposed evidence for tectonic activity on Ganymede's surface. These generated tidal forces, suggested to have been 'pulsed' as Ganymede, Europa, and Io eventually settled into their current resonance over an extended period of time, were also supposedly responsible for heating Ganymede's iron core to the point where the satellite's intrinsic magnetosphere could be generated (something absent in Io and Europa despite both also being subject to the same tidal forces). Currently, evidence of tidal forces at work within Ganymede are now negligible, in stark comparison to Io and Europa, begging the question of what force now sustains Ganymede's supposedly internally heated dynamo that is supposedly currently powering its intrinsic magnetosphere?

Added to the above scenario for the shaping of Ganymede's surface are the assumed various impact and bombardment events from assorted meteorites, asteroids and other debris left over from the primordial Solar System's theorised nebula accretion disk. These are thought to have provided the majority of silicate, mineral and organic materials now locked into the ice of Ganymede's crust. The constant spluttering of Ganymede's surface by charged particles originating from Jupiter's irradiating magnetosphere are also considered to be a surface changing factor and most responsible for releasing gas molecules from the icy surface to form Ganymede's very thin oxygen atmosphere.

There is also the almost unique geological-like concept of cryovolcanism to add into the mix. This is the belief that volcanic-like activity can occur through the upwelling of 'warm ice' through the icy crusts of satellites like Ganymede. Upon reaching the surface, these percolating plumes of warm ice are thought to be partially responsible for the creation of so-called ice calderas, ice geysers, and other associated quasi-volcanic features. However, in relation to Ganymede, this is a very tentative theory, an ad hoc response borrowed from other ad hoc explanations for the observed geysers of steam and water vapour seen on icy satellites like Saturn's Enceladus. Regarding any observable presence of it on Ganymede, cryovolcanism as a whole is still a debatable concept.

> "While strong evidence supports the notion that cryovolcanism can occur on icy satellites, the mechanisms

by which water, with its higher density, can be erupted through lower density ice are only partially understood. . ."[2]

Ganymede shows little in the way of active cryovolcanism, something that could only be fully expected in the presence of tidal forces, of which there are virtually none currently affecting this Jovian satellite. In summary then, the mainstream view of how Ganymede came to look like it looks today is:

"Its surface displays an array of geologic features spanning a wide range of ages, which record evidence of the internal evolution of a large icy satellite, dynamical interactions with the other Galilean satellites, and the evolution of the population of small bodies impacting the surfaces of the satellites."[3]

But does it?

Here we have another example of mainstream scientific consensus struggling to fit the available data into preconceived models. If Ganymede accreted in its current orbit, as the accepted model demands, then current thermal models of planetary formation have no way of explaining why Ganymede *currently* has a hot molten iron core — and that includes factoring in the supposed past effects from discordant orbital-induced tidal forces. Ganymede's currently negligible internal tidal activity should have seen the satellite cool down hundreds of millions, if not billions of years ago to such a degree that it would not be able to generate the supposed heat convection needed to support the internally driven dynamo the consensus says is needed to power its known magnetosphere.

Because of these problems, many of mainstream science's most cherished planetary beliefs actually face almost insurmountable

---

[2] Michael Thomas Bland, *"The Tectonic, Thermal and Magnetic Evolution of Icy Satellites"*, page 20

[3] Geoffrey C. Collins, et al, "Ganymede science questions and future exploration," *Planetary Science Decadal Survey Community White Paper*, PDF, see Introduction section, http://www.lpi.usra.edu/decadal/opag/GanymedeScience.pdf

problems in explaining Ganymede. Add to this the belief that the Galilean moons of Jupiter all supposedly accreted out of the same dusty disk surrounding a primordial Jupiter, yet are radically different in both their compositions and surface features, and you start to understand why we are left with an increasingly confused and illogical picture regards the mainstream consensus view of how the Jovian system formed.

Frozen Remnants of an Antique Solar System

In the previous chapter, we proffered that Ganymede's uniquely low moment of inertia was more likely due to its outer mantel being composed of pumice rather than ice. This was created, we have suggested, by the extreme electrical heating of Ganymede during a flaring of Jupiter as Jupiter underwent a nova-like event. After the subsidence of this flaring event, Ganymede's 'pumiced' outer mantel became an electro-magnet for the accumulation of vapour from Jupiter's huge clouds of water until an ocean rose globally across Ganymede's surface, an ocean populated by the floating shattered shards of pumice split off from the outer mantel's main bulk. Subsequently, Ganymede became a liquid saturated moon orbiting the dull glow of a sub-brown dwarf safely inside the Sun's warming habitable zone.

For how long this persisted, we don't know. Did life come to exist on such a world? . . . Possibly, but we don't know for sure as yet.

However, what we can surmise is that something changed on Ganymede. . . and changed dramatically!

Artisic impression of the surface of Ganymede today. The Pentagon building in Washington DC would fit neatly into the crater in the foreground.

Whatever pushed Jupiter out into its current orbit was responsible for the icy globe that Ganymede has become today. The main culprit, it is suggested by this work, was another free-floating *sub*-brown dwarf ensnared by the Sun. That *sub*-brown dwarf came with its own collection of satellites, the arrival of which triggered massive electrical discharges throughout the now overpopulated Solar System and caused the subsequent catastrophic rearrangement of the Sun's old and new planets into the current order we see today.

The *sub*-brown dwarf responsible for this calamitous re-ordering of the Solar System was destined to become the gas-giant planet called Saturn, and with it came the planets Earth, Mars, Venus, Neptune and Uranus. For the Jovian system of worlds, retreat into the cold outer regions of the Sun's influence spelled doom for any life that may have existed on its four Galilean moons. Gradual or sudden, such a shift inevitably froze over Jupiter's former water worlds while Jupiter itself sputtered spectacularly from *sub*-brown dwarf to gas-giant, incapable of continuing to provide the warming glow so necessary for the prolongation of any possible life on the surfaces of its former liquid-covered satellites. The Ganymede we see today is thus an icy tomb preserving the last vestiges of whatever world may once have existed there during the epoch we have dubbed the Antique Solar System.

However, the dying throes of Jupiter under electrical assault due to the arrival of Saturn would have seen a last flurry of breath-taking planetary electrical activity. These events would have burned themselves into the memory of any who might have witnessed them.[4] Mythology tells us the inhabitants of the newly arrived Earth did indeed witness enhanced electrical activity on and around the planet Jupiter, events that have etched their memory into the collective human psyche in the form of that most awesome of mythological archetypes; the Jovian thunderbolt! One can only imagine what might have been occurring on Ganymede at any time these thunderbolts were unleashed. It is perfectly plausible that Ganymede would have borne the full brunt of Jupiter's dying electrical spasms as the old Antique Solar System gave way to the new.

Do we see evidence of Ganymede having suffered such violence?

The Electrical Scarring of Ganymede

More than evidence for tectonic activity or cryovolcanism, Electric Universe (EU) adherents see copious evidence for the scarring effects of atmospheric electrical discharge on the moons of Jupiter. Europa's surface in particular shows distinct signs of dramatic electrical strikes on a vast scale, far larger than the puny lightning bolts experienced here on Earth.

> "Europa displays a frozen record of strikes by Jupiter's thunderbolts in the recent past. Just as lightning looks for the easiest path to ground, Jupiter's thunderbolts preferred to run across the surface of Europa rather than through the near vacuum of space. The result is a filamentary pattern of

---

[4] E.g., Dwardu Cardona, "Primordial Star", pp 345 - 346

superimposed furrows running this way and that for hundreds and thousands of kilometers across the face of the moon."[5] (See images below)

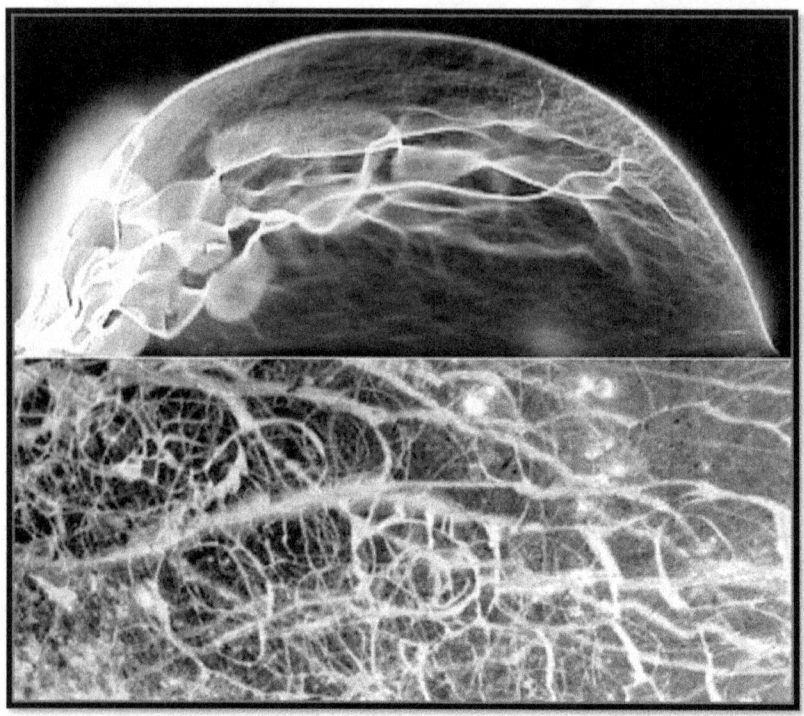

A comparison of electrical filaments in a plasma ball (top) to surface features on Europa (bottom). The icy surface of Europa seems to have perfectly preserved the electrical scars of a Jovian thunderbolt as it passed over its surface. Picture credits: Top image, Wallace Thornhill; bottom, NASA.

The standard take on this type of surface feature is that it is somehow similar to what we see occurring at our own polar ice sheets here on Earth.

---

[5] "Picture of the Day (POTD), December 14, 2005," Thunderbolts.info, see: http://www.thunderbolts.info/tpod/2005/arch05/051214europa.htm

"We also know that it's [Europa is] covered in thousands of cracks that look very much like the type we see in ice floes floating on liquid water here on Earth."[6]

By 'ice floes floating on liquid water here on Earth,' it is assumed what is meant is something like the following image:

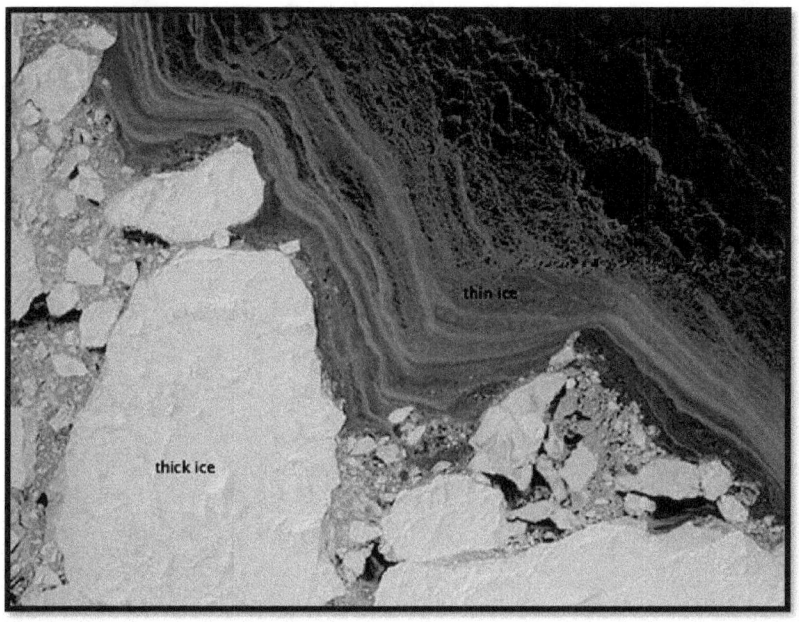

Arctic ice sheet at the ocean's edge. Supposedly a comparative phenomenon to what is seen on Europa. Image credit: NASA.

To actually compare the two supposedly similar phenomena, below is another shot of Europa's own supposed ice floe-like surface features:

---

[6] Phil Plait, "Huge lakes of water may exist under Europa's ice," *discovermagazine.com, Bad Astronomy*, November 17, 2011.

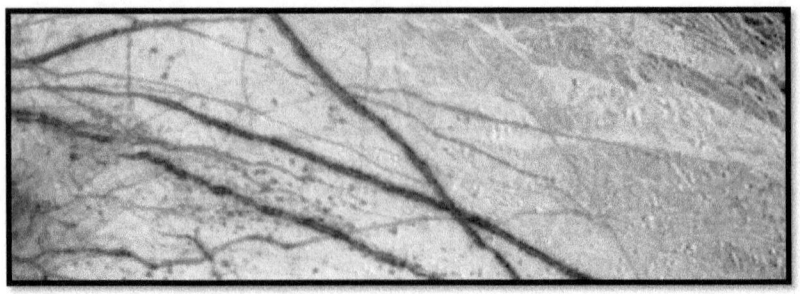

The supposed ice floe-like cracking of Europa. Image credit NASA.

The crisscrossing flowing nature of Europa's dominant and wide dark lines somehow *don't* evoke images of the arctic ice cap, yet this doesn't stop mainstream science from seeing what it wants to see. The colour differentiation should in itself be a warning that all is not well with an Arctic/Europa ice sheet comparison. While we completely agree that it is ice that we are looking at, the bizarre maze of lines seen on the ice of Europa must surely have another explanation to it simply being a form of ice floe cracking (see next image).

Jupiter's moon Europa clearly displays an electrically scarred surface. Image credit NASA.

Ganymede's own surface is testament to similar scarring patterns, but it is the remarkable differences between the lighter and darker parts of Ganymede's terrain that may demonstrate the true nature of destruction suffered by Ganymede in the dying days of the Antique Solar System. The lighter areas of Ganymede are thought by mainstream opinion to have come about as a result of what planetary scientists call resurfacing. Typically, these areas display long groove patterns in stark contrast to the more crater sputtered dark areas.

> "Heavily cratered dark terrain, which has similarities to the surface of Callisto, covers one third of the surface of Ganymede. The other two thirds of Ganymede has been resurfaced to form light terrain, much of which has been tectonically modified by structures known as "grooves."

> Whether the primary mode of resurfacing is tectonic or
> cryovolcanic is still an outstanding question."[7]

So, if, as according to the above quote, the mode of resurfacing is 'still an outstanding question,' then why are we then confidently assured that much of the light terrain has been 'tectonically modified'? Is there a chance that other forces may have been at work in carving out these distinctly differently shaded areas of Ganymede?

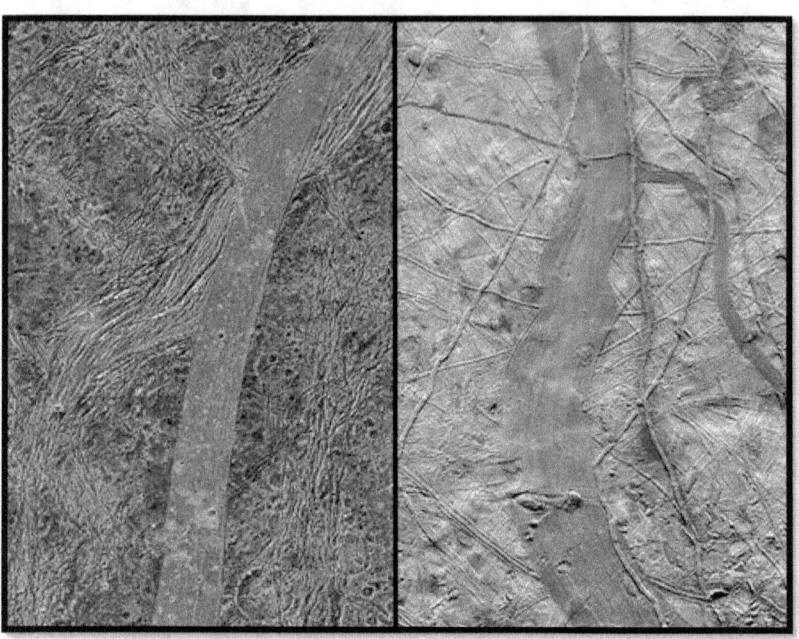

"This image, taken by NASA's Galileo spacecraft, shows a same-scale comparison between Arbela Sulcus on Jupiter's moon Ganymede (left) and an unnamed band on another Jovian moon, Europa (right). Arbela Sulcus is one of the smoothest lanes of bright terrain identified on Ganymede, and shows very subtle striations along its length. Arbela contrasts markedly from the surrounding

---

[7] Geoffrey C. Collins, et al, "Ganymede science questions and future exploration," *Planetary Science Decadal Survey Community White Paper*, PDF, see Surface Geology section,
http://www.lpi.usra.edu/decadal/opag/GanymedeScience.pdf

heavily cratered dark terrain." Exert from original caption released with image: Image Credit: NASA/JPL/Brown University

The similarity of features seen in the preceding image clearly shows that both Ganymede and Europa were subjected to similar surface shaping forces, yet we are told that both moons are remarkably different in internal composition, with Ganymede being 'highly differentiated' compared to Europa. On Ganymede it is thought that its crust is completely decoupled from its outer ice mantel by a liquid ocean. This would, therefore, provide a huge buffer to any heating or tectonic forces deriving from the Ganymede's deeper silicate outer core. Accordingly, if Ganymede's internal composition is so different to that of Europa (Europa's surface not being decoupled from its outer mantel), then how is it that both moons are affected the same way by supposedly the same internally driven tectonic forces?

Electrical scaring on a cosmic interplanetary scale offers a far better explanation for the surface features we see etched in the ice of the Jovian moons today. This fits neatly with the theory that these two moon's host planet, Jupiter, was catastrophically disrupted and moved at some point in its history, subsequently subjecting both moons to similar electrical discharges on a massive scale. In this way we do not need to reconcile the same observable surface results on entirely differently composed satellites by appealing to disassociated internal tectonic forces.

While mainstream consensus assures us that Ganymede's surface features are comparable to those seen on Earth, they are simply stumped by some of the other seemingly inexplicable features found on Jupiter's largest satellite. For example, the following image has geologists at odds on how to explain the nature of the 'tectonic' features seen when compared to known geological processes observed on Earth.

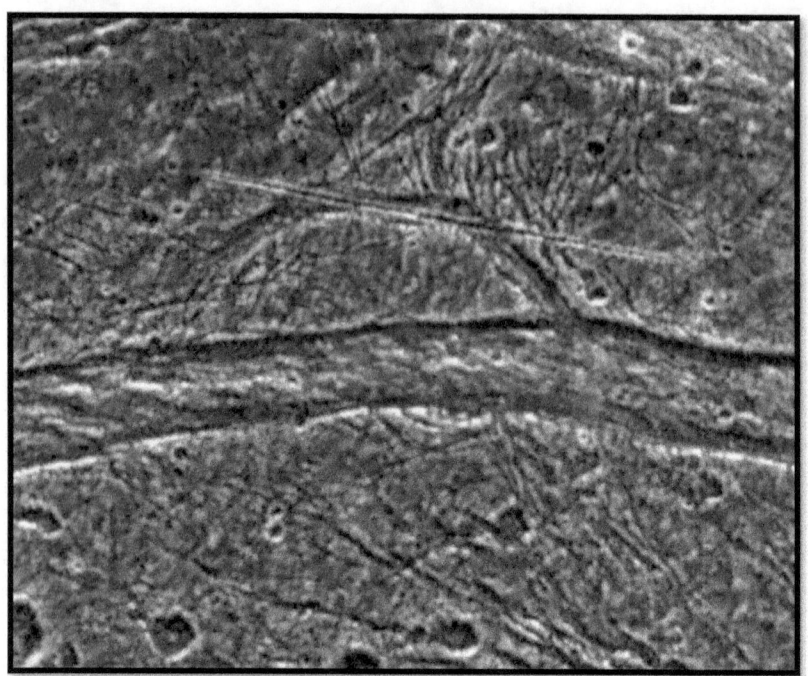

This part of Ganymede contains many complex supposed tectonic structures, with small fractures crisscrossing the image. The image covers an area approximately 80 kilometres (50 miles) by 52 kilometres (32 miles) across. Image credit: NASA/JPL/Brown University

At issue amongst geologists in the above image is the almost perfectly straight 'fracture' crossing the looping 'fault' line that seems to branch out from a major so-called fault line bisecting the image from left to right.

"This image is centered on an unusual semicircular structure about 33 kilometers (20 miles) across. A 38 kilometer (24 miles) long, remarkably linear feature cuts across its northern extent, and a wide east-west fault system marks its southern boundary. *The origin of these features is the subject of much debate among scientists analyzing the data.* Was the arcuate structure part of a larger feature? Is the straight lineament the result of internal or external processes? Scientists continue to study this data in order to

220

understand the surface processes occurring on this complex satellite."[8] (Emphasis ours)

What should be kept in mind when viewing these images of Ganymede's surface is that the predominant material involved is ice. From this we are told that certain features are a result of ice flowing like lava and that these 'ice lava flows' are what form some of the more interesting regions of contrast seen on Ganymede. Like some kind of perfectly straight glacier, these enormous rivers of ice are said to have originated from under Ganymede's crust to emerge through the uncertain process of cryovolcanism. Once on the surface, the upwelling warm ice from the icy outer mantel below then supposedly began its long and often very straight journeys across Ganymede's surface.

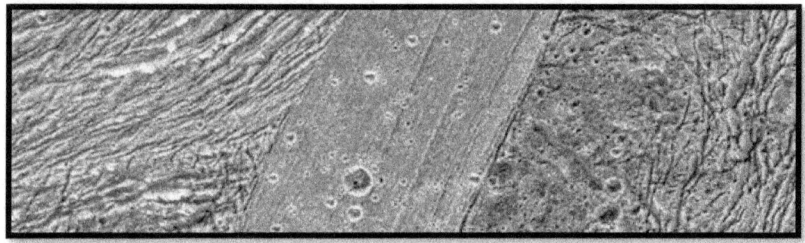

Close-up of the Nicholson Regio and Arbela Sulcus area of Ganymede shows the stark contrasts between many of the diverse terrain types found on the satellite's surface, all believed to be caused by tectonic forces, or by straight-flowing so-called ice lava. Credit: NASA/JPL.

[8] Jet Propulsion Laboratory, Photojournal, "PIA01087: Geological mysteries on Ganymede," exert from original caption released with image, NASA/JPL/Brown University, see http://photojournal.jpl.nasa.gov/catalog/PIA01087

For example, in the following image from Ganymede's surface released by the Jet Propulsion Laboratory it is said that:

> "The relatively smooth appearance of Sippar Sulcus [bottom portion of the image] hints that icy volcanism once paved over the area."[9]

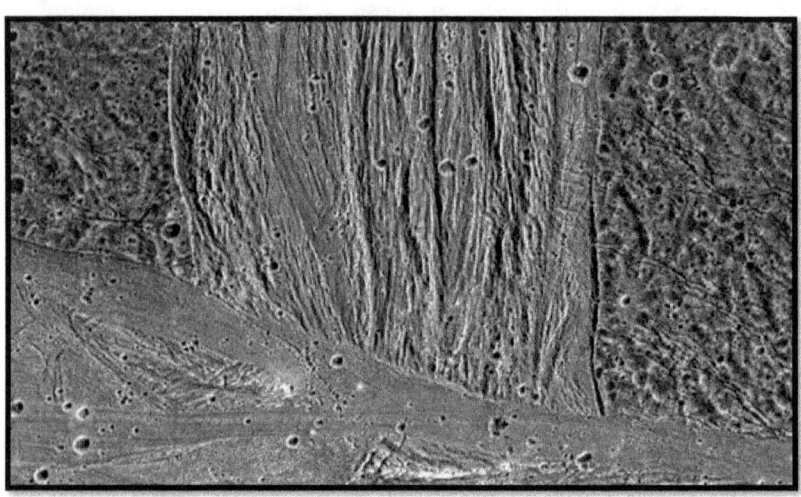

The truncated termination of the long-grooved Erech Sulcus region by the brighter, smoother terrain of the Sippar Sulcus (bottom), the latter a supposed example of cryovolcanic ice lava flow re-paving over Ganymede's darker regions.

As noted previously, the mechanics for so-called cryovolcanism are only partially understood at best. What is often not stated is that there are no clear origin points on Ganymede for these supposed ice lava flows; no obvious volcanic craters or calderas have been positively identified as producing these ice lava flows. Yet, we are somehow expected to believe this mechanism can produce ice lava flows capable of re-paving sections of Ganymede up to hundreds of

---

[9] Ibid, see image "PIA01615: Swaths of Grooved Terrain on Ganymede", http://photojournal.jpl.nasa.gov/catalog/PIA01615

kilometres long. Surely, once on the surface, this upwelling of 'warm ice' would be as much subject to the chilling temperatures of Ganymede's exposed topside as the rest of the ice covering the surface. So how did it continue to flow in its supposed warm state for hundreds of kilometres? Why didn't it just build up into mounds like we see molten rock lavas doing on Earth, and even on Ganymede's sister moon Io?

Other explanations for the image seen on Ganymede's surface have led to it being compared to fault lines seen in the East African Rift where the land is being pulled apart by Earth's own tectonic forces. It's believed a similar tectonic process (during the theorised period of settling into the Laplace resonance) pulled apart areas of Ganymede's crust and the ensuing rift was filled in with warmer and much fresher ice from below. Again, we are struck by the straight lines involved in this proposed tectonic process on Ganymede, yet a look at the East African Rift region from space shows the comparison to be strained at best.

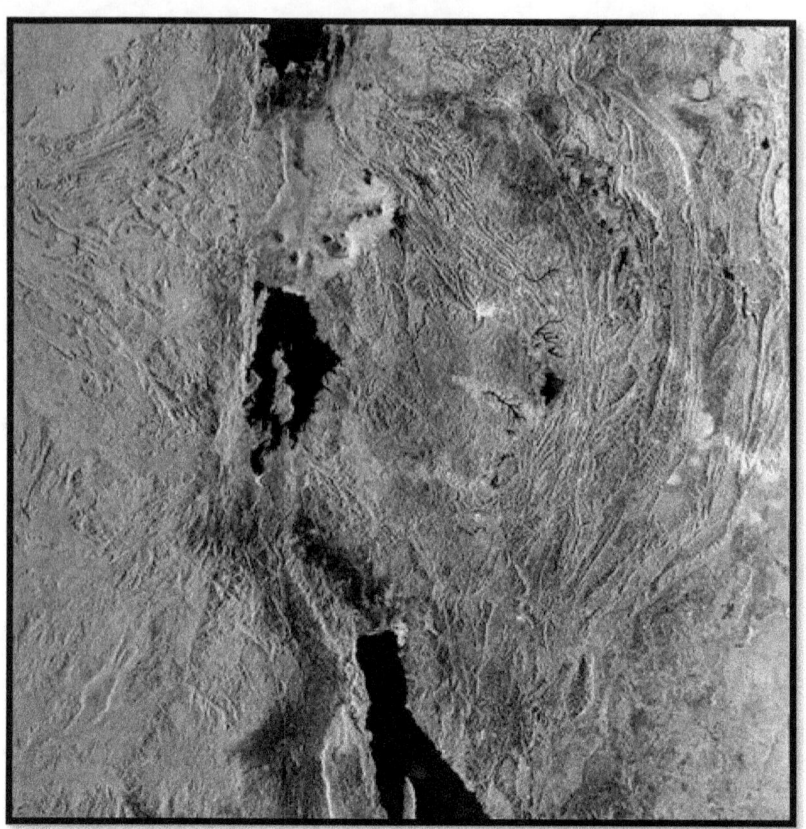

The lakes region of the East African Rift. The distinctive straight-edged demarcations between regions as seen on Ganymede are completely lacking here. Image credit: European Space Agency.

The comparison is further strained because; on the one hand we are dealing with ice on Ganymede, and on the other silicate-based land on Earth. Comparisons between the behaviour of two completely different materials should be viewed as suspect to start with, yet the above view of the East African Rift from space clearly emphasises the marked differences in these two geologically shaped surfaces from different parts of the Solar System.

At this point in the narrative enough has been said to cast doubt over conventional geological theories regards the formation of Ganymede's distinctive ice surface features. Sudden and catastrophic freezing of Ganymede's global ocean followed by an immense amount of scarring from electrical discharge remains our preferred explanation for what came to shape Ganymede's frozen crust of water. As stated, Ganymede's icy crust is believed to have been formed when Jupiter was removed from the Sun's habitable zone, along with its satellites, to its current frigid orbit. That Ganymede's surface is a preserved testament to extreme electrical events at this time also points to the sheer destructiveness of Jupiter's displacement. If so, then, what are we to make of Ganymede's surface as evidence for it having once been a warm water world populated with pumice bergs and rafts and possible organic life?

In the following chapter we will speculate on what kind of water world Ganymede had been during the Antique Solar System epoch and look to see if there is evidence buried in its icy surface alluding to this time. What we will find is that, not only is there significant evidence for fixed and floating land features on Ganymede, but there are strong indications that organic substances inhabited these land forms. We will find that, frozen in the vast ice crust tomb of Ganymede's surface, there are the tell-tale signs that life itself once possibly teemed across Ganymede's surface, both above and below the surface of its global ocean.

Summary and Takeaways from this chapter

In this chapter we have questioned the standard geologic, tectonic, cryovolcanic, meteor impact, convection and tidal explanations for the surface features observed on the ice encrusted Jovian moons. An electrical explanation is offered instead, especially for both Europa and Ganymede, in which the striated and pock-marked surface regions of both moons are attributed to intense electric discharge events coming from an electrically active Jupiter at the close of the Antique Solar System epoch, a period that saw Jupiter catastrophically displaced by the arrival of Saturn outwards to its current wider and colder orbit.

- The evidence for dramatic electrical scarring on the surfaces of both Ganymede and Europa suggests a catastrophic end to the once liquid worlds of ancient Jupiter's moons. The rapid cooling and subsequent icing over of these moons served to capture and preserve the effects of destructive electrical discharges in the shaping of their surfaces as seen today.

- The accretion model for the formation and evolution of the Jovian moons is discredited in light of the increasingly contradictory data concerning the internal and thermal properties of these satellites. Current observations simply do not support the consensus model of internal mechanisms being the foremost forces in the formation of Ganymede and Europa's icy surface features — especially in the case of Ganymede. Electrical forces and extreme discharge events are a far better explanation for the diverse internal compositions and surface features found on these very different, yet very close Jovian neighbours.

- On closer inspection, the icy surface features of both Ganymede and Europa in no way correspond at the macro level in either shape or form to the ice cracking found at Earth's polar regions, nor do they correspond to Earth's tectonic rift regions. The concept of cryovolcanism as a mechanism for some of the ice features found on Ganymede is to be treated with caution due to no actual observations of this hypothesised thermal process having been recorded; it remains to be fully understood, if at all ever understood.

- While it is accepted that Ganymede, Europa and Callisto offer completely different internal compositions, mainstream consensus fails to explain why the same tectonic and cryovolcanic forces are said to occur on such completely different composited bodies. Again, cosmic scale electrical scarring of the Jovian moons' rapidly iced-over surfaces, caused by Jupiter's catastrophic displacement at the end of the Antique Solar System epoch, is our favoured alternative explanation for the shared surface features found on the differently composited ice moons of Jupiter.

- The thick icy crusts of Ganymede and the other Jovian ice moons serve to entomb the changes and consequences of whatever catastrophic events overtook these satellites in their ancient past. Had these moons, especially Ganymede, once been liquid water worlds, then it is expected that their now ice encrusted surfaces will have preserved evidence of their former warm liquid states pointing to the distinct possibility that life once existed on these bodies.

- It is suggested that evidence for concentrations of pumice-supported organic materials will be preserved in the icy depths of what was once Ganymede's global liquid ocean, another leading indicator that life may have once thrived on this moon.

Club Ganymede: The Antique Solar System's Tropical Paradise

Science fiction heavyweight Arthur C. Clarke wrote three sequels to the hugely successful *2001: A Space Odyssey* in which an alien species is depicted as having ignited Jupiter into a star that then transforms the moon Europa into an oceanic tropical paradise. This is done ostensibly to help in the evolution of a new sentient species called the *Europans*. However, the possibility that a Jovian moon could have provided the conditions for harbouring life in the distant past has led us in previous chapters of this work to focus on Ganymede as the most likely contender for a similar scenario — minus the god-like aliens and their evolutionist projects.

In building a case for Jupiter as having once been a sub-brown dwarf that orbited the Sun within its habitable zone, we have also discussed how Ganymede's unique possession of a protective magnetosphere, plus its probable liquid ocean (now an ice crust) and its indications of a past oxygen-based atmosphere could have provided the ideal habitat for life as we know it. We have addressed the issue of Ganymede's perplexingly low moment of inertia and come to the conclusion that the satellite's outer 800 kilometre thick-plus mantel is not a massively deep, salty and icy ocean, as is currently thought, but one made of pumice, the result of the super-heated electrical baking of the entire moon during a nova-like flare event in Jupiter's distant past.

What this has left us with is a speculative picture of a warmed Ganymedian surface covered in a liquid global ocean, the depth of which may have been anywhere between 15 – 40 kilometres deep. Such a worldwide ocean would likely have been dotted with the floating fractured shards of pumice that would have been split off from its outer pumice mantel. These pumice fragments would then possibly provide Ganymede with variable floating land masses that we will now refer to as *pumice bergs*.

While it is entirely plausible that there would be some peaks jutting up from the pumice mantel itself, it is likely that these would be rare fixed islands in a sea of giant floating pumice bergs, the latter of which would eventually drift together to form haphazard rafts and archipelagos. With these in place, the existence of organic compounds in the form of electrically induced *tholins* (formed from methane) would mix and mingle with pumice dust and form the basis for soil and organic growth on Ganymede.

Given the above scenario, an intrepid interplanetary traveller visiting Ganymede at this time could have then laid out a deck chair and basked in the glow of a two-sun world while watching a whole ecology get to work *aqua-forming* this intriguing little ball of wet pumice. Whatever the process, the surface of Ganymede would have come to resemble a rich tropical water world complete with its own climate and uniquely fragile islands populated with various flora and fauna.

A Day on Ganymede

Ganymede's orbit around Jupiter currently takes approximately 7.1 days to complete. There is no reason to suggest that this was not the case during the Antique Solar System epoch, though a different orbital time is entirely possible. If we stick with the current orbital duration for Ganymede, then we can make some observations as to what life on Ganymede could have been like within this postulated two-sun cosmos.

The abundance of light striking Ganymede from both Jupiter and the Sun during our proposed Antique Solar System would undoubtedly make for an all-round illuminated existence — at least on Ganymede's Jupiter-facing hemisphere. Ganymede's orbit around Jupiter is phase-locked like our own moon is to Earth, i.e., the same side of Ganymede always points towards Jupiter. Any vegetation or creatures living on the Jupiter-facing side of Ganymede could have expected some form of light at all times, even during that time when the Sun would have been entirely and overwhelmingly occulted by Jupiter. At these times, a fainter yet perceptible reddish-to-purplish glow would still shed an energising

229

red-spectrum light on this side of Ganymede, as the below graphic demonstrates:

A topographical cartoon view of antique Jupiter from above its north pole with Ganymede represented at the four main stations of its phase-lock orbit around Jupiter. The bright effect of the Sun's light can be seen striking Ganymede from left of frame, while Jupiter's purplish sub-brown dwarf glow dully illuminates Ganymede's Jupiter-facing hemisphere. Only when Ganymede is behind Jupiter would a full hemisphere (the outward-facing hemisphere) experience total night. Ganymede currently takes approximately 7.16 days to complete a full orbit of Jupiter. Image not to scale.

However, for any plants or creatures living on the away-facing side of Ganymede, life would have been drenched in unforgiving ultraviolet sunlight for a good portion of the 7.1 days it takes Ganymede to circle Jupiter. On this side, the Sun would have described a slow arc across the sky of Ganymede's outwardly-facing hemisphere, taking about three and a half days to complete its journey before slowly setting into a long dusk, followed by a deep night, and then the eventual emergence into a long dawn. As already noted, the whole process would take 7.1 Earth days.

What can be deduced from this two-sun effect radiating on Ganymede as it journeyed around a glowing Jupiter is that, depending on where you were on Ganymede, any flora and fauna would experience variations and mixtures of the red-light and ultraviolet light spectrums. On a world where most life could be expected to be of the marine variety, the filtering effects of Ganymede's ocean depths would easily shield it from the Sun's more deadly ultraviolet radiation, where needed, just as our own oceans do on Earth. For any terrestrial life that may have established itself on the tops of our postulated pumice bergs, life directly under the Sun would have been potentially bright and harsh, while those periods spent exclusively in Jupiter's red-light spectrum would have approached something similar to a semi-permanent twilight existence.

Also, the drifting nature of our proposed pumice bergs could see any terrestrial life eventually experience all the light extremes that Ganymede's two suns had to offer. Only those potentially rare fixed peaks of pumice jutting up from the ocean floor could expect to receive a regular and predictable regimen of light similar to that experienced on land here on Earth. Had these fixed islands found themselves on the outward side of Ganymede, long exposure to the Sun would probably see them become bleached desert islands, along with any pumice bergs that may have run aground there. If they were located anywhere else on Ganymede, such peaks would become prime real estate for any species of terrestrial creatures or vegetation Ganymede could produce.

There is, however, another possibility to be factored into this scenario of drifting pumice bergs and that has to do with the thermal heating the Sun would have generated to produce currents in Ganymede's global ocean. There is reason to believe that a substantial portion of these proposed pumice bergs and rafts would eventually collect together and come to dominate one hemisphere. There they would form a buoyant landmass interlaced with complex myriads of watery channels, lagoons and shifting pumice swamplands.

The light reaching Ganymede from two suns would undoubtedly have been the single largest factor contributing to Ganymede's environment, its climate and any ecology that grew there. There are of course other factors, such as any internal volcanic activity from Ganymede's interior and any radiated internal heat. These important points will also be discussed in a later section of this chapter, but for now we should turn to the current evidence we have concerning Ganymede's surface and see if these can indicate what kind of world a tropically warm Ganymede might have been.

Desert Islands: Rocky Lumps and Minerals in the ice

It took scientists looking at the *Galileo* space probe's images of Ganymede nearly seven years to announce in 2004 that something odd could be detected in the thick icy crust that encapsulates the satellite:

> "Scientists have discovered irregular lumps beneath the icy surface of Jupiter's largest moon, Ganymede. These irregular masses may be rock formations, supported by Ganymede's icy shell for billions of years."[1]

These rock formations are tentatively believed to be suspended in the thickness of the ice surrounding Ganymede. They are not thought to have a solid base of contact with Ganymede's supposed icy mantle due to the theorised liquid ocean sandwiched between the mantle and the icy crust, though some think they may be piles of rocks reaching down that far.

---

[1] "Scientists Discover Ganymede Has A Lumpy Interior," *Science Daily*, August 17, 2004, see:
http://www.sciencedaily.com/releases/2004/08/040817082023.htm

"The findings have caused scientists to rethink what the interior of Ganymede might contain. The reported bulges reside in the interior, and there are no visible surface features associated with them. This tells scientists that the ice is probably strong enough, at least near the surface, to support these possible rock masses from sinking to the bottom of the ice for billions of years. But this anomaly could also be caused by piles of rock at the bottom of the ice."[2]

What can be ascertained is that, whatever these rock formations are, it is unlikely they are a collection of meteorite or asteroid impact debris due to the said lack of 'visible surface features associated with them.' How such a mass of rubble and rock got to be collected and situated as a pile within or just under Ganymede's 'icy shell' is a genuine problem for the standard model pointing to Ganymede's differentiated interior, and even its supposedly impacted exterior surface.

"The anomalies could be large concentrations of rock at or underneath the ice surface. They could also be in a layer of mixed ice and rock below the surface with variations in the amount of rock," said Dr. John Anderson, a scientist and the paper's lead author at JPL. "If there is a liquid water ocean inside Ganymede's outer ice layer there might be variations in its depth with piles of rock at the ocean bottom. There could be topographic variations in a hidden rocky surface underlying a deep outer icy shell. There are many possibilities, and we need to do more studies."[3]

Given our own scenario, such a finding poses no problem in understanding where these "concentrations of rock" originate. Unburdened with the necessity of needing a vast liquid ocean under Ganymede's crust to account for the Jovian satellite's low moment of inertia, these ice-bound rock anomalies become instead the potential coagulations of the free-floating pumice bergs postulated

---

[2] Ibid.

[3] Ibid.

in the first section of this chapter. That they may actually constitute the peaks of fixed land masses jutting up from a pumice-based outer mantle is also in keeping with our scenario and the *Galileo* data.

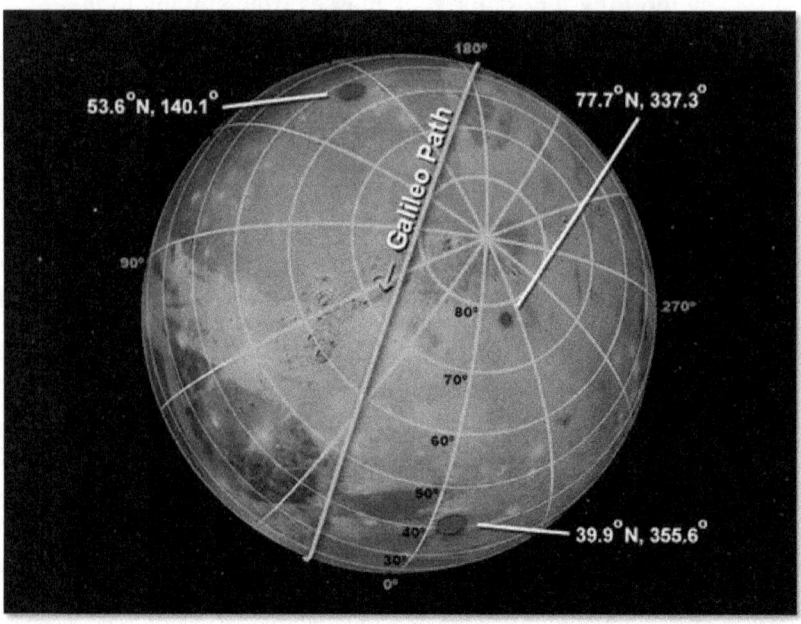

Ganymede's odd patches of interior rocky lumps shown in red. One of them is almost the size of Switzerland. According to the original caption with this image: "Scientists have discovered irregular lumps beneath the icy surface of Jupiter's largest moon, Ganymede. These irregular masses may be rock formations, supported by Ganymede's icy shell for billions of years." Image credit: NASA/JPL.

Then there are the images captured by the Galileo probe's Near Infrared Mapping Spectrometer (NIMS), which can be compared to the Voyager probe's image of Ganymede's surface to show just how much water does cover Ganymede — and where all the minerals locked in its icy crust are. There is an interesting correlation between the concentrations of minerals locked in Ganymede's ice and the darker areas where there seems to be fewer ice concentrations on the moon's surface. Minerals are often associated with silicate-based materials leading to the suspicion that the areas of high mineral content on Ganymede may, in fact, be harbouring collections of mineral-rich silicate materials. How they all managed to congregate into distinct areas in Ganymede's icy crust is a question the mainstream model for Ganymede's evolution fails to answer.

However, an ancient Ganymede awash in a global liquid ocean could be expected to develop large concentrations of minerals that find themselves afloat in and on its swirling currents. We believe it most likely that the highest proportion of these minerals originated in the left-over dust of Ganymede's electrically baked pumice mantel. There will be a proportion of minerals that arrived with impacts by the usual assortment of meteorites, comets, etc., but these are considered negligible when interpreting the majority of so-called impact craters as actually being the results of electrical strikes suffered by Ganymede during the Solar System's transition from its antique epoch to its current manifestation. In fact, even to this day Ganymede gives off a significant amount of 'dust,' another indication that an immense heating event struck the satellite at some time in its past.

> "Ganymede is a significant dust source, implying that its surface composition could be sampled directly from orbit."[4]

One wonders just where all this dust come from and what event might have put it into space?

---

[4] R. T. Pappalardo, K. K. Khurana, and W. B. Moore, "THE GRANDEUR OF GANYMEDE: SUGGESTED GOALS FOR AN ORBITER MISSION," Brown University (Dept. Geological Sciences, Providence RI 02912-1846; 2UCLA (Los Angeles, CA 90095-1567), 4065.pdf, see: http://www.lpi.usra.edu/meetings/outerplanets2001/pdf/4065.pdf

The image of Ganymede at left was taken by the Voyager probe, while the other images come from the Galileo mission's NIMS camera. The green indicates water coverage with the brighter green representing more water. The image on the right indicates concentrations of minerals (red speckles) in Ganymede's ice, which, interestingly, correspond with the darker patches where less water is observed — a possible indication for land-type concentration? Image credit: NASA/JPL.

The Dust of Land and Life: Clay and Organic Compounds

The presence of clays and organic materials in Ganymede's darker regions are pause for thought for anyone contemplating the existence of life on Jupiter's largest moon. Because mainstream science considers Ganymede's surface to have experienced its current frigid climate for millions, if not billions of years, the existence of clays and organic materials is explained as having arrived during impact events from the very same space debris that Ganymede supposedly accreted from.

"Ganymede's dark terrain material contains clays and organic materials that may indicate the composition of the impactors from which jovian [*sic*] satellites accreted."[5]

What is not explained is why these 'impactors' have produced more dust around Ganymede than virtually any other highly impacted moon in the Solar System. After all, Ganymede's sister moon Callisto has received just as many so-called impacts, if not more, yet there is no dust cloud matching that which we see around Ganymede. As noted previously, this dust cloud could, theoretically, be sampled from orbit, a curious state of affairs for a planet supposedly locked in a massively deep layer of ice for billions of years. Given our pumice hypothesis, it seems more likely this dust is the desiccated remnants of electrically vaporised pumice ejected into orbit as a result of intense electrical strikes. Dust clouds are commonly observed phenomena where electrical discharges contact solid surfaces.

While readers thus far will be aware of our complete rejecting of 'accretion' theory for planetary birthing, we do however recognise that clays and organic materials must surely have had their origins in whatever material constituted Ganymede at its birth. The Electric Universe (EU) model allows for extreme electrical discharges to take place between celestial bodies during times of electrical imbalances, a scenario we believe may have been played out between Ganymede and Jupiter during the births of the Jovian moons Io and Europa. During such a phase electrical discharges could quite easily account for the emergence of electrically transmuted elements displaying organic qualities.

Having already previously discussed how such an event may have resulted in Ganymede's outer silicate-based mantel having been electrically baked and transformed into a deep layer of pumice, we have also commented on the expected layers of dust and ash that would coat such a surface in the aftermath of such an event. Any further, but lesser electrical activity striking Ganymede could see

---

[5] Ibid.

chemical compounds such as methane mix with pumice dust to form *tholins*, a form of organic compound that can act as a base for further organic development. It is most likely that at least some form of tholin-like compound accounts for the detection of Ganymede's own organic compounds.

Tholins are a heteropolymer molecule thought to be formed by the solar ultraviolet irradiation of simple organic compounds such as methane or ethane. They do not occur naturally on Earth due to the high oxygen count here, which inhibits the process, but they are currently found in great abundance on most of the icy moons of the outer Solar System — moons like Ganymede. Of course, mainstream science, with its rejection of the notion of electrical charge being carried in space, can only conceive of tholins being a product of sunlight striking oxygen deficient planetary bodies. However, laboratory experiments in which tholins have been successfully reproduced have to use electrical discharge to substitute for ultraviolet irradiation for their creation.[6] It is not difficult to see how a moon such as Ganymede, with its copious evidence of vast electrical scarring (discussed in previous chapter), could find itself inundated with tholins and the beginnings of a soil-based ecosystem — long before the accumulation of an oxygen-rich atmosphere.

However, we also need to consider the flip side to the presence of tholins on Ganymede; and this is that they more likely represent the burnt-off remains of carbon-based organic compounds that have been *tholin-ised* by the subsequent and rapid electrified destruction of whatever oxygen-rich atmosphere Ganymede may have once enjoyed!

---

[6] "The term "tholin" was coined by astronomer Carl Sagan and his colleague Bishun Khare to describe the difficult-to-characterize substances he obtained in his Urey-Miller-type experiments on the gas mixtures that are found in Titan's atmosphere. It is not a specific compound but is a term generally used to describe the reddish, organic component of planetary surfaces." Exert from Wikipedia entry on 'Tholins', which cites: Carl Sagan & B. N. Khare (1979). "Tholins: organic chemistry of interstellar grains and gas". *Nature* 277 (5692): 102. Bibcode: 1979Natur.277..102S. doi:10.1038/277102a0

Tholins offer an effective screen against the harmful effects of ultraviolet radiation for any planetary surfaces they might cover, pretty much in the same way that vegetation does on Earth. In this way tholins are not dissimilar to the broken down decomposed remains of carbon-based vegetation, which provide layers of organic film (oil and coal deposits) within Earth's crust and on its surface. The main difference between the two compounds is that tholins seemingly can only be formed in an oxygen-deficient environment, while carbon-based organic deposits (called PAHs)[7] require oxygen to form. Otherwise, they both form via a burning-off process, of which electrical discharge can be but one of many. Having already noted that tholins have been produced in the laboratory by the use of electrical discharge, their presence on Ganymede's electrically scarred surface seems almost self-explanatory — they are the electrically burned-off residue of whatever complex organic compounds once populated that now icy and oxygen-deficient moon.

But tholins, we are told, can only form in the absence of oxygen, while carbon-based organic compounds are the signature of an existing oxygen-rich environment. Here is a paradox; how do you go from oxygen-produced carbon-based organic compounds to tholins created in a non-oxygen environment? The solution might seem almost trite in its simplicity; during the breaking down process you get rid of the oxygen — *immediately!* Which brings us to the concept that Ganymede's tholins are the result of electrically degraded carbon-based organic compounds being chemically transformed by the catastrophic and sudden loss of that moon's oxygen-rich atmosphere, a dynamic we will discuss in more depth shortly.

In the meantime, while evolutionists have long speculated that tholin-type compounds provided the initial environment for the earliest advent of life as we know it, we, on the other hand, believe

---

[7] Polycyclic aromatic hydrocarbons — PAHs are organic compounds found oil, coal, and tar deposits, and are produced as byproducts of fuel burning (whether fossil fuel or biomass), a process that needs oxygen.

this is putting the cart before the horse. As said, tholins on Ganymede are most likely the result of the electrical baking of existing life and not the precursor to life. Experiments show that intense electrical discharges can readily break down and purify organically contaminated bodies of water by changing complex organic compounds into species not dissimilar in structure to tholins.[8] Ganymede seems to have been home to an organically contaminated body of water, but because it has eluded mainstream scientists that electrical discharges rained down on Ganymede at some time in its past, they have not been able to see the origins of Ganymede's organic compounds for the proverbial carbon-based woods.

We put it that the tholins detected by scientists on Ganymede's surface could simply be the electrically decimated and decomposed remnants of more complex organic compounds, the immolated remains of organic matter destroyed, and not created, by intense electrical and plasma discharges into Ganymede's once liquid water environment.[9] Such massive and destructive discharges would also explain the sudden loss of Ganymede's once oxygen-rich atmosphere, a necessary component to the emergence of tholins, and a subject that we will now address.

---

[8] See for example experiments in the degradation of the compound phenal (a form of PAH) in water when subjected to electric discharges: Anto Tri Sugiarto, Masayuki Sato, "Pulsed plasma processing of organic compounds in aqueous solution," *Department of Biological and Chemical Engineering, Faculty of Engineering, Gunma Uni_ersity, 1-5-1 Tenjin-cho, Kiryu-shi,Gunma 376-8515, Japa,* Received 17 June 1999; accepted 12 July 2000

[9] The labelling of organic compounds detected on Ganymede as 'tholins' is purely down to that satellite's current known lack of an substantial oxygen atmosphere. Had there been an existing oxygen-rich atmosphere on Ganymede, then mainstream science would undoubtedly be trumpeting the discovery of extra-terrestrial carbon-based organic compounds and musing on the implications for extra-terrestrial life within our Solar System.

Postcards from Ganymede's Lost Atmosphere and Climate

Can a moon the size of Ganymede support a life-sustaining atmosphere?

This question is best answered by taking a look at another satellite orbiting a gas-giant in the Solar System; the Saturnian moon called Titan. Saturn's most interesting ice-moon has defied efforts to fully map its surface due to it being cloaked in a hazy, thick atmosphere that is more dense at its surface than Earth's own atmosphere; 1.45 times as dense[10] in fact. Titan even has its own giant and seemingly perpetual hurricane situated at its south pole.[11] Despite holding no oxygen, Titan's ability to support a large and dense atmosphere settles the issue in the affirmative of whether ice-moons like Ganymede could have ever sustained atmospheres capable of supporting life as we know it.

So what can this tell us about Ganymede's atmosphere?

Had Ganymede once supported a dense oxygen atmosphere capable of sustaining life, then life on Ganymede's outward-facing hemisphere would have been exposed the longest to the Sun's damaging, yet tholin-friendly ultraviolet rays. However, there would eventually and inevitably have been the filtering effects of clouds and precipitation as water molecules evaporated into the atmosphere from Ganymede's liquid ocean, just as it is experienced here on Earth. But to have precipitation, and even liquid water for that matter, it goes without saying that Ganymede must have had a substantial atmosphere capable of supporting water vapour clouds.[12]

---

[10] "Titan (moon)," Wikipedia entry, see
http://en.wikipedia.org/wiki/Titan_(moon)#Atmosphere

[11] In the Electric Universe model Tatan;s south polar hurricane would be generated by a Birkeland current.

[12] Absent an atmosphere, water will go from its solid state (ice) to its gaseous state (water vapour) without becoming a liquid in the interim.

At present, Ganymede's atmosphere is near non-existent, though it does sport a whiff of oxygen and ozone.

> "Astronomers using the Hubble Space Telescope found evidence of thin oxygen atmosphere on Ganymede in 1996. The atmosphere is far too thin to support life as we know it."[13]

So, did Ganymede lose its atmosphere, or did it never possess a significant atmosphere in the first place?

True-color image of layers of haze in Titan's extremely dense

---

[13] "Ganymede: Overview," *Solar System Exploration* website, NASA, see: http://solarsystem.nasa.gov/planets/profile.cfm?Object=Jup_Ganymede

nitrogen-rich atmosphere. There is no oxygen in the atmosphere of this moon of Saturn, making it an ideal place for the existence of the organic compounds called tholins. The fact that Titan sustains such a heavy atmosphere is proof Ganymede is too capable of supporting a significant atmosphere. Image credit: NASA

Mainstream science's answer to this question is predictable — Ganymede's atmosphere is as it is today because that is the way it has evolved. This, of course, presupposes the uniformitarian belief that Ganymede has virtually always existed in the same type of celestial environment as it does today and that the key to its past is to simply extrapolate backwards over aeons from the processes we can observe taking place in its current orbital environment. The thin presence of oxygen on Ganymede is thus thought to merely be the result of ionised particles sputtering the surface and releasing limited amounts of oxygen molecules from the H2O ice compound that makes up Ganymede's icy crust.[14]

However, we would argue that Ganymede's current thin oxygen atmosphere (really its exosphere) is a last vestige of a former oxygen-rich and dense atmosphere, which was burnt off during Jupiter's electrical outbursts, events that have left their mark on Ganymede's surface and that were clearly witnessed by ancient mankind who recorded them in the world mythological record.[15] The vast electrical strikes that scar Ganymede's icy surface today are testament to this explosive end to Ganymede's atmosphere as a significant gaseous body, an event that would have taken place at the closing of the Antique Solar System epoch with the arrival of the Saturnian system of planets. Such strikes would effectively

---

[14] See: D. T. Hall, P. D. Feldman, M. A. McGrath, and D. F. Strobel, "The Far-Ultraviolet Oxygen Airglow of Europa and Ganymede," *The Astrophysical Journal*, Volume 499, Number 1, (1998), doi:10.1086/305604

[15] See: "Is Lightning the Strongest Creative Force?," Picture of the Day, *Thunderbolts.info*, July 16, 2009, http://www.thunderbolts.info/tpod/2009/arch09/090716strongest.htm

have burnt off any existing oxygen and reduced Ganymede's atmosphere to the levels we see today.

These strikes, particularly those whose points of contact are preserved in the huge so-called Tashmetum and Hershef craters, were the result of huge *atmospheric* electrical discharge events. Like lightning strikes on Earth on a giant scale, the Jovian thunderbolts responsible for these two so-called craters would have needed to *pass through an atmosphere* to produce the effects we now see etched into Ganymede's surface today. It bears repeating; electrical discharges capable of rendering the giant radial marks seen on Ganymede would have needed to pass through an *atmosphere* to produce such an effect.

Why?

The answer is found in what is known in the electrical engineering field as *dielectric breakdown*.[16] A dielectric is a substance that can be used to separate two electrical charges to form what is called a capacitor. Keeping those two charges separate is what stops them from discharging or sparking. Glass, plastics, and certain oxidized metals make excellent dielectrics. So does dry air.

On a planetary scale, an atmosphere acts as a giant dielectric between a negatively charged planetary surface and positively charged space. This forms a massive capacitor, which, when overloaded by extreme electrical input (a Jovian thunderbolt for example), will break down and set off massive electrical discharge events. The subsequent huge lightning strike will scar the negatively charged planetary surface with a distinctive Lichtenberg pattern;[17] the very same type of patterns we see emanating out of Ganymede's Tashmetum and Hershef craters (see image below).

---

[16] For an explanation on how dielectric breakdown manifests itself as lightning here on Earth, see: "Dielectric Breakdown," *Picture of the Day*, December 31, 2010, Thunderbolts.info, http://www.thunderbolts.info/tpod/2010/arch10/101231breakdown.htm

[17] See: "Lichtenberg Figure," *Wikipedia* entry; http://en.wikipedia.org/wiki/Lichtenberg_figure

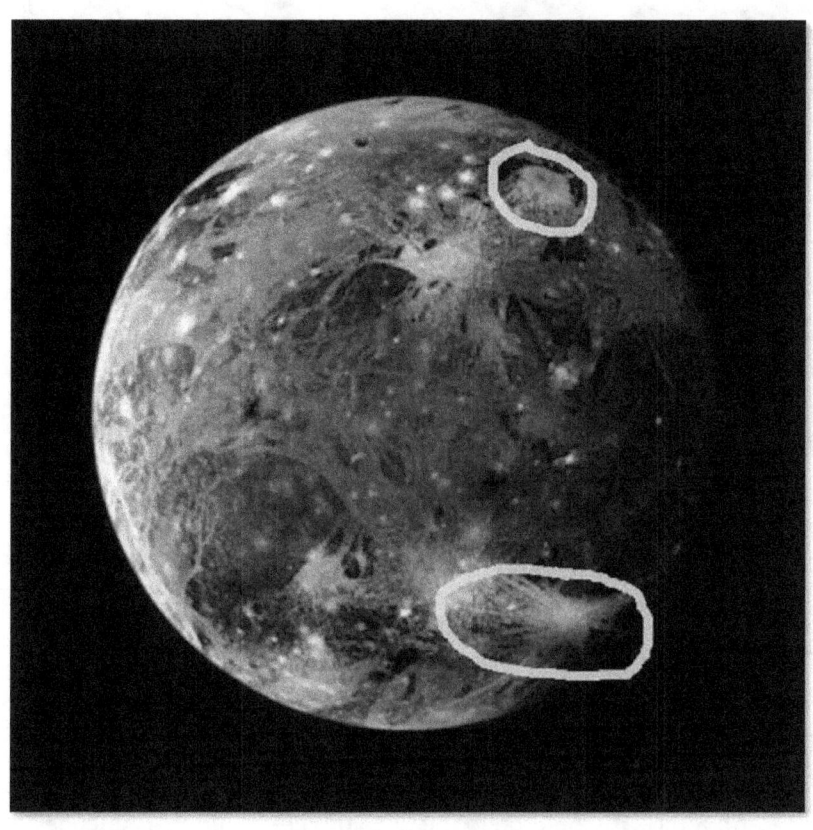

*The Hershef crater (ringed white radial pattern at upper right) and
the extreme radial rays of the Tashmetum crater (ringed lower right)
on Ganymede are better understood as contact points for vast
electrical discharges powered by Jupiter in ancient times and
triggered by the entry of the wandering brown dwarf Saturn into the
Sun's electrical domain. Each strike region is as large as the
Kingdom of Spain. A thick atmosphere would have been present on
Ganymede to facilitate these huge electrical strikes through
dielectric breakdown. These Jovian thunderbolts were witnessed by
humans and recorded in world mythology. Image credit: NASA.*

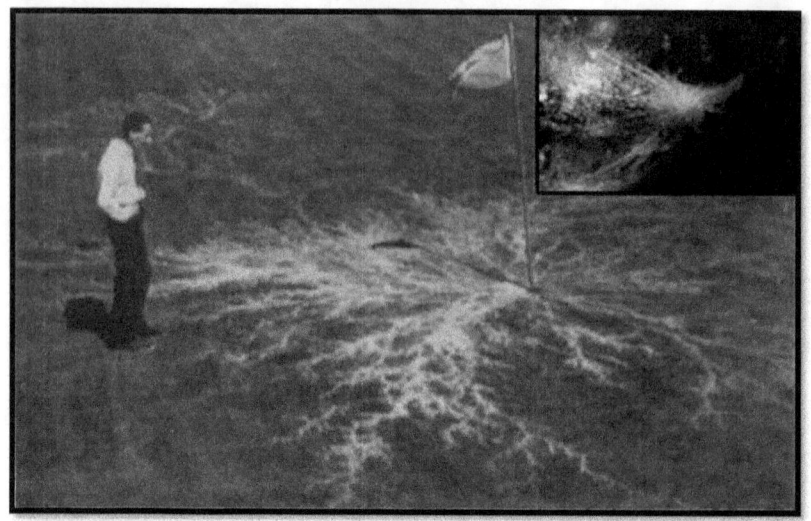

The distinctive Lichtenberg pattern left after a lightning strike on a golf course. Compare to the Tashmetum crater on Ganymede (insert), an example of the results of electrical discharges taking place at planetary scales.

Given the presupposition that Ganymede did indeed once support an oxygen-rich atmosphere capable of sustaining the kind of life we see on Earth today, it would be to the oceanic meteorological sciences that we would look to next for any insights into the actual climate enjoyed on Ganymede during the Antique Solar System epoch. The combination of the Sun's and Jupiter's different degrees of warming would have set up competing thermally-driven jet streams in its atmosphere and warming currents in its global ocean provoking storms every bit as powerful and destructive as seen on Earth. Precipitation could have been extreme under the thermal conditions afforded by the heat of two suns and Ganymede's polar regions would undoubtedly have had ice caps every bit as foreboding as those found on Earth.

Regards Ganymede's landscapes or seascapes as it may, it is the presence of these proposed thermal currents and thermally-driven atmospheric turbulences that would have seen Ganymede's pumice archipelagos gradually drift about its oceanic surface. Ganymede would have been a world undergoing constant change to its buoyantly fragile landscape as its vastly huge pumice bergs and film-like pumice rafts milled about in the oceanic currents. Any vegetation clinging to any floating pumice would have been predominantly green in hue, though parts of Ganymede's Jupiter-facing hemisphere would have been populated with a mixture of reddish and green vegetation wherever Jupiter's more red-dominated light spectrum struck its surface. And, the three and a half days of night experienced on Ganymede's outwardly-facing hemisphere would not have been too cold due to Jupiter's substantial plasma sheath reflecting a portion of the sub-brown dwarf's energy back inwards.

In short, the climate on Ganymede would have been ideal for life as we know it here on Earth, albeit a more marine oriented form of life capable of thriving in Ganymede's oceanic environment. It is to Ganymede's global ocean that we now turn to complete our postulated picture of Ganymede as a liquid tropical paradise.

Water, Water Everywhere… and Plenty to Drink

Something briefly overlooked in the previous section of this chapter, but mentioned elsewhere, is the fact that liquids of any type on a planetary surface need the pressure of an existing atmosphere to sustain them in their liquid form. Any solid such as ice will immediately move into its gaseous form upon reaching its melting point in the absence of a substantial atmosphere. So it follows that for Ganymede to have once enjoyed a liquid ocean on its surface there must have been a substantial atmosphere present in the first place. This, we have argued in the above, was indeed the case at some point in Ganymede's distant past, a stance attested to by the abundant evidence of atmospheric electrical discharge strikes scarring Ganymede's icy surface today.

However, the current discussion in mainstream circles concerning the presence of a liquid ocean on Ganymede centres on the belief that one is locked away under Ganymede's airless and thick icy crust, and it is this issue that we must address first before we can continue on with our proposition of a warm liquid ocean having once covered Ganymede's surface.

The existence of a salty ocean on a foreign planetary body always gets astrobiologists and evolutionists somewhat flushed. This is because it is believed that life evolved out of the Earth's oceans and, therefore, it may have done the same thing elsewhere in the universe where salty oceans are found. A salty ocean is particularly alluring because of its electrical conductivity, a factor recognised in biology as the key component to the functioning of life within organic organisms. Humans, the highest known carbon-based life form, are mostly made up of salty water, which facilitates the biological electrical energies that flow the essence of life and consciousness through all of us. Evolutionists believe that this harkens back to our distant emergence from the primordial soup of the Earth's salty oceans, so they go looking for the same environmental conditions on other worlds.

There is an irony that astrobiologists and evolutionists would recognise the role played by electricity in the emergence of life, yet their astrophysicist colleagues almost completely reject its role in the formation of solar systems and galaxies. Where they do converge on the issue of Ganymede is in their belief that the supposed deep ocean hidden under Ganymede's icy crust is salty. As noted in previous chapters, this belief is based on the need to explain the satellite's conductivity in maintaining an intrinsic magnetosphere supposedly generated from deep within its highly differentiated interior. A deep salty ocean is said to provide this conductivity for the electricity coming up from some supposed internal dynamo that then powers Ganymede's magnetosphere.[18]

---

[18] The real cause of Ganymede's intrinsic magnetosphere has been discussed at length in the chapter entitled "Ganymede: Third Rock from Jupiter". Also put forward is the hypothesis that Ganymede's highly differentiated interior and low moment of inertia is the result of its outer mantel being made of pumice and not icy salt water.

There is, however, a problem with this — Ganymede seems to be remarkably deficient in the element Sodium; i.e., salt.[19]

Sodium is thirteen times less abundant around Ganymede than it is around the other Jovian moon Europa. Io, Jupiter's highly volcanic moon, is awash in sodium. However, Ganymede is not, and whatever small amounts of sodium have been detected on its surface is thought to have somehow found its way there from Ganymede's supposedly deep salty ocean.[20] This leaves accepted models of Ganymede's interior composition dependent on all Ganymede's salt being locked away under its icy crust, an icy crust that appears to be a body of relatively fresh water.

Of course, readers who have made it this far will know that we postulate an entirely different internal composition for Ganymede that replaces its supposed deep salty ocean and icy outer mantle with one composed of pumice, leaving its icy fresh water crust as the satellite's true predominant global body of water.

The implications of the above statement are astounding in light of claims that Ganymede once enjoyed a warm climate with its now icy crust having once been a liquid ocean. Ganymede may have been entirely unique; a **fresh water** world where saline water would have been the exception and not the normal.

It is only fair to point out that current data regarding Ganymede's sodium content is still extremely thin. We will not actually know if Ganymede's icy crust has a large sodium contingent until we actually visit its surface. But for now it appears Ganymede is

---

[19] By 'salt' we refer to sodium in its various manifestations as what is called 'rock salt' and not to sodium chloride. Sodium is an alkali metal found in various minerals such as rock salt and is the sixth most abundant element in the Earth's crust. A lack of sodium in Ganymede's atmosphere is blamed on its magnetosphere which may be fending off energetic particles capable of creating the chemical transformations to produce sodium. Or it may simply be that Ganymede lacks sodium per sé.

[20] See; Michael E. Brown, "A Search for a Sodium Atmosphere around Ganymede," *Icarus, Volume 126, Issue 1, March 1997,* Pages 236-238.

largely sodium deficient relative to its sister moons and the Earth. This is definitely a problem for mainstream evolutionist thinking regarding the emergence of life on Ganymede, let alone for traditional solutions to the existence of its magnetosphere. However, a fresh water ocean and oxygen-rich atmosphere on Ganymede may be exactly what we are looking for in terms of the *sustaining* of organic organisms and life as we know it — especially higher forms of life . . . Hmmm!

The Gravity of Life on Ganymede

Ganymede's gravity is calculated to be about $1/7^{th}$ that of our own planet; is there any reason to believe that would have created any insuperable problems for humans or other creatures living on Ganymede in past ages?

The answer would appear to be no. We've already seen that Earth's own gravity in the age of dinosaurs could not plausibly have been more than about a third of its present value and it was probably less than that. There is the added consideration the creatures, and that includes humans, living in water would get sufficient exercise moving their own bodies against water (i.e. swimming), particularly in the case of humans who, lacking tails and fins, require more energy to swim then do fish or aquatic mammals. There is no reason to believe that humans living under such circumstances would require weightlifting in order to stay in shape.

For that matter, and recalling the comment (end of the chapter on ancient gravity) about human back problems being due to the strong gravity of our present world, the much weaker gravity of Ganymede appears to add to the picture of Ganymede as an ideal world for humans.

The only real question as to gravity in such a scenario would be that of the lesser-gravity world having a breathable atmosphere and we've already covered at least most of that question; there are also possibilities of atmospheres of such worlds being held together by electrostatic or electromagnetic forces rather than entirely by gravity, and there is the possibility that Al DeGrazia raised of the atmosphere within the planetary system of a dwarf star being general to the system rather than separate per the individual planets.[21]

Behold! A Tropical Splendor in the Ancient Heaveens

Glorious Ganymede: A fresh water world during the Antique Solar System epoch when Jupiter orbited the Sun in its habitable zone. This is Ganymede's trailing darker hemisphere. In this interpretation of Ganymede's surface just before its freezing over, green vegetation has colonized a floating landmass at lower right made up of collected pumice bergs while other smaller landmasses can be seen to the north. The dark material in the blue water areas are heavier submerged pumice bergs while the sandy colored patches are thin filmy-like pumice rafts. The green landmass is riddled with dark swampy lakes, lagoons and estuaries. Virtually all land floats atop Ganymede's global fresh water ocean. The northern polar ice cap can be seen at this angle. Artist's impression by the authors.

---

[21] Alfred DeGrazia, "Solaria Binaria", p. 62 - 68

The world that was Ganymede during the epoch we have dubbed the Antique Solar System seems to have been a uniquely fresh water world populated by vast, free-floating pumice bergs coagulating into tenuous and fragile landmasses covered in organic matter and possible vegetation. Warmed and lighted by both the Sun's ultraviolet rays and Jupiter's own red-spectrum light, Ganymede would have been more than favourable for any variety of complex organic compounds, and even carbon-based organisms. Any human stepping back into such and age would have found living on Ganymede quite comfortable, albeit a little more wet and weightless than here on Earth.

Ganymede's Jupiter-facing hemisphere during the Sun's occulting behind Jupiter. The red-light spectrum given off by Jupiter as a sub-brown dwarf would have cast a reddish purple hue over Ganymede. Any vegetation on this side could be expected to take on a predominantly reddish hue. One of Ganymede's anomalous rocky lumps can be seen as a purplish patch at the top of the coagulated pumice landmass now known as the Galileo Regio area. The southern hemisphere is mostly populated by giant, yet flimsy and thin pumice rafts. Artist's impression by the authors.

The presence of fresh water would have boded well for higher life forms and contributed significantly to a robust ecosystem involving flora and fauna suited to seashore, lakefront, swamp, and estuary conditions similar to that found on Earth. A key feature of such a world would be its abundance of littoral zones, areas where Ganymede's ubiquitous ocean would form shores and lagoons amongst the archipelagos of our postulated collections of pumice bergs. Littoral feeding on sea foods and aquatic fruits would be the main constituent in any diet supporting any possible higher fauna in the absence of carnivorous predation. The closest approximation on Earth to this scenario would have been the primordial shores and islands of Lake Victoria and the swampy flooded hinterlands of the Amazon basin.

This chapter and those preceding it in Part II of this work have sought to offer a case for the existence of a habitable planetary body in our solar system during a time before Earth had yet arrived. We offer the Jovian moon of Ganymede as the most likely contender for this planetary body and express our confidence that it was once indeed a world bursting with the ingredients for life.

*Size comparison between Ganymede and Earth with Ganymede
presented as it might have appeared during the postulated Antique
Solar System epoch. Earth image courtesy of NASA.*

But alas, such a world was never to last and today Ganymede is a
frigid ice-encrusted and scarred remnant of its former self, a stark
testimony to the fragile nature of existence in this universe. Should
further exploration confirm that life did once exist on Ganymede,
then it is to world mythology that we must turn to piece together the
circumstances of its demise. For it is noted that in the epic
interplanetary cataclysms recorded by ancient mankind the power of
the Jovian thunderbolt was one of the most feared cosmic weapons
stalking the celestial realms of the old gods. Closer to the action
than anyone today would think possible, mankind witnessed these
events, these horrors, and recorded them in a myriad of ways for
future generations. Yet, one wonders if the sight of Jupiter's wrath
descending on its third and largest moon might have provoked
fearful realisation; that in the demise of an almost forgotten oceanic
paradise there was thankfulness in having been spared the sharing of
such a fate.

Did mankind on Earth possibly witness the destruction of their most ancient home and the legacy of their origins left behind on another world?

Summary and Takeaways from this chapter

In this chapter we have speculated on the type of environment Ganymede would have enjoyed had it been a liquid water world orbiting Jupiter in its ancient phase as a glowing sub-brown dwarf, itself orbiting the Sun within the habitable zone. We have presented a picture of a *bright* world enjoying the benefit of both the Sun's and Jupiter's warming influences, a world with a global ocean and an oxygen-rich atmosphere where pumice bergs would have formed free-floating land concentrations perfect for the sustaining of aquatic-based life.

- Due to the phase-lock nature of its 7.1 day orbit around Jupiter, Ganymede's oceanic surface would have experienced different types of light at different regular intervals in different hemispheres. Only the outward hemisphere would have experienced the contrast of night and day as we experience it here on Earth, while the Jupiter facing hemisphere would have received blended degrees of both the Sun's and Jupiter's solar warmth.
- Enigmatic, yet sizable lumps of rock concentrations have been detected within the icy crust of Ganymede, suggesting that they are suspended there and supported by the ice itself. These rock concentrations are inexplicable according to the standard model of Ganymede's composition, and do not conform to the idea that they are collections of impactor fragments. Their existence does, however, support the notion of concentrations of free-floating pumice bergs coagulating in the currents of a once watery world.
- Concentrated areas of mineral deposits in Ganymede's icy crust suggest further evidence of coagulation taking place due to the ebb and flow of oceanic currents — they do not support the idea of their arrival by impactors on an existing ice crust where the static nature of a global ice sheet would preserve a more randomly dispersed collection of mineral deposits.

- The detection of clay and organic compounds in the ice of Ganymede again points to the possibility Ganymede may have once supported life. The extreme cloud of dust that surrounds Ganymede points to the true source of these clays and organic compounds. This cloud is thought to be the leftovers of impact events on Ganymede, yet Callisto's supposedly equally impacted past has produced no such cloud and Ganymede remains almost unique in the vast amount of dust collected around it. We suggest this dense cloud is the vaporized residue leftover from Ganymede's electrical baking during the formation of its pumice-like mantel.

- The existence on Ganymede of organic compounds called tholins is thought by mainstream scientists to point to the initial development of potentially complex organic compounds by the effects of solar radiation — a precursor to the emergence of life according to evolutionary theory. We, on the other hand, argue that their presence points to them being evidence of the rapid destruction of existing organic compounds, and even existing carbon-based compounds, by atmospheric electrical discharges taking place on a planetary scale.

- The claim for atmospheric electrical discharges having taking place on Ganymede is evidenced by the existence giant radial Lichtenberg-like contact scars observed at many points on Ganymede's surface. Erroneously referred to as 'impact craters', these massive electrical contact scars are proof that Ganymede once enjoyed a thick atmosphere. This is due to an atmosphere being necessary for the *dielectric breakdown* process in producing an atmospheric discharge event.

- Ganymede's currently thin *exosphere* indicates that a previously thick atmosphere would have been rich in oxygen. This thicker oxygen-rich atmosphere would have been burned off by the catastrophic electrical discharge events as evidenced by the giant Lichtenberg-like scars referred to in the previous point. Ganymede's previously thick atmosphere was a casualty of Jupiter's catastrophic displacement on the Sun's capture of Saturn at the end of the Antique Solar System epoch.

- While most atmosphere-supporting planets and moons in the solar system tend to have predominantly carbon-dioxide or methane-based atmospheres, Ganymede joins Earth and Europa as one of the few Earth-like places in our solar system with oxygen as a main component of their atmospheres. This augurs well for the idea that Ganymede should be considered as the one known place beyond Earth most likely to support or have supported life as we know it.

- Ganymede and its icy crust are remarkably deficient in salt, a fact that indicates that its former global liquid ocean would have been uniquely freshwater-based.

- Ganymede's low gravity would have posed no problems to aquatic-based life. Even humans, who could adapt to an aquatic lifestyle on a warm water Ganymede, would experience no detrimental effects due to the lack of an Earth-like gravity.

- Ganymede during our postulated Antique Solar System epoch would have been a tropical-like fresh water world fully capable of sustaining life at all levels, including human life. The evidence for Ganymede's ability to have once provided the type of conditions necessary for the support of a wide diversity of aquatic-based fauna and flora makes Ganymede the most likely contender in our current solar system for the discovery of evidence for life as we know it beyond Earth.

Stellar distances, Friar Occam, and Ganymede

In the "Caveats" section (caveat 6) at the beginning of this book we mentioned that:

> In the absence of time machines, this work makes heavy use of the logical principle called "Occam's razor." Named after Friar William of Occam, the principle is generally understood to mean that of competing theories with equal explanatory power, the simplest should be preferred. In particular, given the immense distances between stars in our galaxy, in the presence of a completely plausible origin for modern man within our own solar system, theories involving saltations from other star systems are ruled out.

What we have seen in the preceding sections of this book, is that Jupiter's moon Ganymede would have been a perfect world for Elaine Morgan's Aquatic Ape hypothesis, and that it would have also been the sort of bright world for which humans would be well adapted.

The other possibilities that somebody might wish to entertain would have modern man arising on something entirely **LIKE** Ganymede, which no longer exists, or arriving on Earth from the vastness of interstellar space. Either of those two possibilities would have to be viewed as a probabilistic miracle, or zero–probability event.

John Cameron's "Avatar" was the first science fiction movie ever to give viewers a realistic idea of how great the distances between stars actually are; the film showed a spaceship using hypothetical antimatter engines taking six years to get from Earth to the nearest star, Alpha Centauri. Another way to look at it would be that if you were to scale our own solar system to having a diameter of about a yard, that is, for the diameter of Pluto's orbit to be about a yard wide, then Alpha Centauri would be slightly more than 4 miles away at that scale; the Sun would be about the width of a human hair at that scale and Earth would be an inch or two away from the Sun.

The basic reality is that for one star to ever capture another unrelated star, i.e., a star that was not involved in any sort of a formation relationship as is the case with Herbig-Haro objects, would be a probabilistic miracle.

Given those realities as well as what we have seen in other chapters of this book, Occam's principle will insist that we view Ganymede as the original home of the human species.

Cro-Magnons and Bible Antediluvians

Compared to that of other animal species, the genetic variation of the human race is so low as to require at least one "population bottleneck" to explain it.[1] All living human beings are said to have had a single male and female ancestor within the last hundred thousand years. Our own general view is that there are at least two basic human groups on the planet and that the differences between them involve their original cultures and technologies while having nothing to do with physical features or anything that one might call 'race'; both groups appear to be capable of producing the myriad of colors or physical attributes found amongst present humans. The two groups we speak of are the Cro-Magnons and their descendants, and the people descended from the biblical Antediluvians.

The term "Cro-Magnon" has, in fact, fallen into general disuse amongst academics due to a confusion as to who or which groups might qualify for membership. Part of the problem arises from scholars viewing recent/gracile hominids like the Skhul/Qafzeh hominids as the earliest examples of modern humans. It also seems clear that some of the American Indian groups, for which we have examples of self-portrait sculptures, did not look entirely like those known Cro-Magnon self-portraits in our possession today. It would be difficult to identify whether these physical differences between these ancient peoples are due to genetic drift or to separate saltations, nonetheless, we feel comfortable in claiming that Cro-Magnon humans and the subsequent familiar antediluvian people of Genesis do amount to separate saltations originating from a common origin.

There is a crucial list of attributes that Biblical and Jewish literature would have to have known about if you wanted to believe that Adam and Eve were directly descended from the Cro-Magnons of

---

[1] See for instance W. Amos and J. I. Hoffman, "Proceedings of the Royal Society, Oct 7, 2009, Evidence that two main bottleneck events shaped modern human genetic diversity"

our postulated purple dawn era, attributes that the Bible and Midrashim appear to know nothing about:

- Stone tools. Adam and Eve and their descendants were metal-tech people from day one (e.g., Genesis 4:22). By contrast, human groups descended from Cro-Magnons went on using stone tools until forced out of it by neighbors descended from Bible Antediluvians if they ever stopped at all. Aztecs were using stone knives and weapons when the Spanish arrived even though they used metals for ornaments and purposes other than tools.
- The atlatl, the signature weapon of Cro Magnon people. Biblical accounts know nothing of this weapon; known biblical weapons are the sword, the spear, the sling, and the bow. By contrast, the Aztecs, a people descended from Cro-Magnons[2], were using atlatls against the Spanish while native Australians continue to use it to hunt kangaroos today.
- Neanderthals and/or other hominids. The Bible knows nothing of hominids. By contrast, the Basque "Basajuan" appears to be a reflection of the Neanderthal and the Australian "Yowie" appears to be some sort of hominid, which native Australians remember. Amerinds likewise in their oral traditions remember dealing with hominids[3] as well as dinosaurs and ice-age megafauna.
- An upper Paleolithic world war in which all hominids were exterminated from the planet, or at least from the parts of the planet that humans inhabit in any numbers.
- A primordial era generically referred to as the "Purple Dawn"; that is, a protracted age devoid of the daylight that we experience today. This period of global gloom and eternal twilight is reported as having been prior to the classical "Golden Age" of ancient literature. Dwardu Cardona, one of the foremost Saturn Theory researchers,

---

[2] Similarities between Salutrean and Clovis Point (and hence Amerind) technologies have been known for some time. One flavour of the thinking connecting French Cromagnon sociegties with Clovis point and Amerind prehistory:
http://frontiers-of-anthropology.blogspot.com/2012/03/solutrean-tools-in-chesapeake-bay-on.html
[3]Ed Fusch, 'S'CWENE'YTI and the stick Indians of the Colevilles'
http://www.bigfootencounters.com/biology/fusch.htm

notes that the Genesis account mentions the creation of light before the sun and moon are created and that the beginning is called an age when "the earth was without form, and void, and darkness was upon the face of the deep."[4] However, Adam and Eve and the Bible antediluvians did not live on Earth during this darkened age and the biblical account beyond Genesis 1:2 does not speak of it again (with the possible exception of Jeremiah 4: 23). By contrast, native Australians refer to this purple-hued age as a "dream time" which their traditions recall vividly, and Amerind oral traditions speak of it as well. As noted elsewhere in this work, the "Purple Dawn" was an age in which daylight as we experience it today did not exist and such light as there actually was during this period was canted towards the blue and red parts of the light spectrum.

Cro-Magnon hunting technology is one of the clearest demarcations separating Cro Magnons from the Biblical antediluvians. The neat thing about Cro-Magnon hunting weapons is that you don't have to guess what they looked like or how they worked. The videos found at the following URLS will perfectly illustrate their effectiveness as hunting tools:

http://www.youtube.com/watch?v=wwNbwPVFkSc

http://www.youtube.com/watch?v=9DDHxOqFkAs

Pity the Winchester, Remington, or Beretta salesman who needed to sell a shotgun to the one gentleman filmed hunting fruit bats (I.e., nobody who uses a boomerang that well *needs* a shotgun):

---

[4] E.g., Dwardu Cardona, "Darkness and the Deep", http://www.bearfabrique.org/Catastrophism/Saturn/drkness.txt

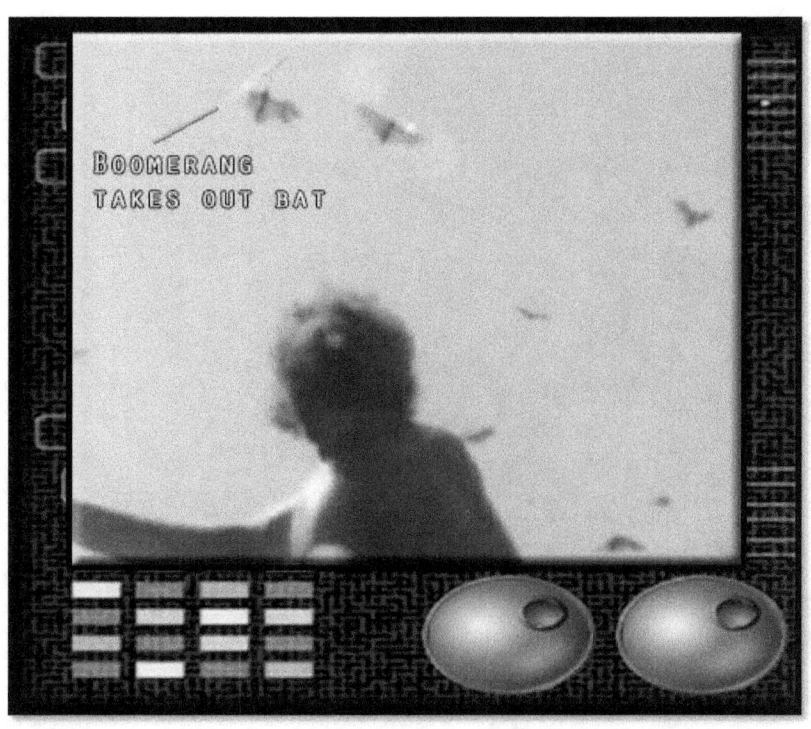

Cro-Magnon hunting technology at work. An Australian Aborigine hunter uses his boomerang to hunt bats in flight. Still shot taken from the film 'Ultime grida dalla savana' (1975)

Likewise, images available on the Internet indicate that proficiency with the atlatl (or "woomera" in the outback) is not limited to one or two specialist exponents of the art or to anthropologists studying these ancient weapons. They still retain their place amongst many peoples enjoying a pre-industrial existence.

The Atlatl: The world's first human super weapon. (image public domain)

Again, if you wanted to believe that Adam, Eve, and the familiar people from the Old Testament were descended from Cro-Magnons, Biblical and Jewish literature would have to know something about the atlatl. This weapon was not just another type of spear; it was the world's first super weapon. If you have ever watched a particularly hard serve in professional tennis, you understand how the atlatl works. A player like Roscoe Tanner or Andy Roddick is basically trading three inches worth of wrist motion for four feet of racket head motion in the blink of an eye, sending a tennis ball off at a blistering 150 mph. An atlatl, if properly used, amounts to doing the same thing with a six foot spear instead of a tennis ball. This weapon would have been the decisive edge in any struggle against either hominids or prey animals that would otherwise not have been possible for humans to hunt. This would especially be the case with competing and predatory hominids, such as Vendramini's Neanderthal (once again shown in the below comparative image).

Given that the Neanderthal made his final European stand in Spain, you'd expect the Spanish Basque, one of the oldest, if not the oldest of human groups in Europe, to remember him . . . and they do. The term "Basajaun" means "forest lord," and the descriptions of these "forest lords" found in traditional literature indicate they were hominids, an observation that entertains the distinct possibility they may have been leftover Neanderthals. Google image searches on "basajaun" turn up pictures that are not that much different from Vendramini's version of the Neanderthal.

At left Danny Vendramini's fearsome reconstruction of the Neanderthal as a predatory hominid, compared to the Basque legend of the 'Basajuan' or 'Forest lord' (at right). Neanderthal image courtesy of themandus.org.

The Bible, however, knows nothing of any of these things. Jewish literature as a whole, to our knowledge, knows nothing of them and, for that matter, Greco Roman literature knows nothing of them either. In other words, people claiming descent from Adam and

Noah seem to have never had to deal with hominids, probably because people of the more ancient Cro-Magnon saltation had essentially eliminated the problem before the arrival of the biblical Antediluvians.

Schemes for dating ancient events such as the arrival of Cro-Magnon people on this planet involve assumptions of uniformity; nonetheless it remains reasonable to believe that some very large expanse of time transpired between the arrival of Cro-Magnon man on Earth, and the time of Adam and Eve's appearance as accounted for in the biblical record. Without a time machine, the best we can offer in this regard amounts to a conjecture. Nonetheless, on a scale of 1 to 10 for conjectures, we view what follows to be a 9.5 as a likely sequence of events:

We have seen that the Saturn system spiraled upwards through space to a point of near contact where its plasma sheath brushed against our present sun's electrically positive heliosphere, and then spiraled back out, an event that was repeated several times before Saturn's outright capture by the sun.[5]

---

[5] See previous chapter "Mankind's Purple Dawn", sub-heading "SATURN AS MASTER OF A FREE-FLOATING PLANETARY NEBULA".

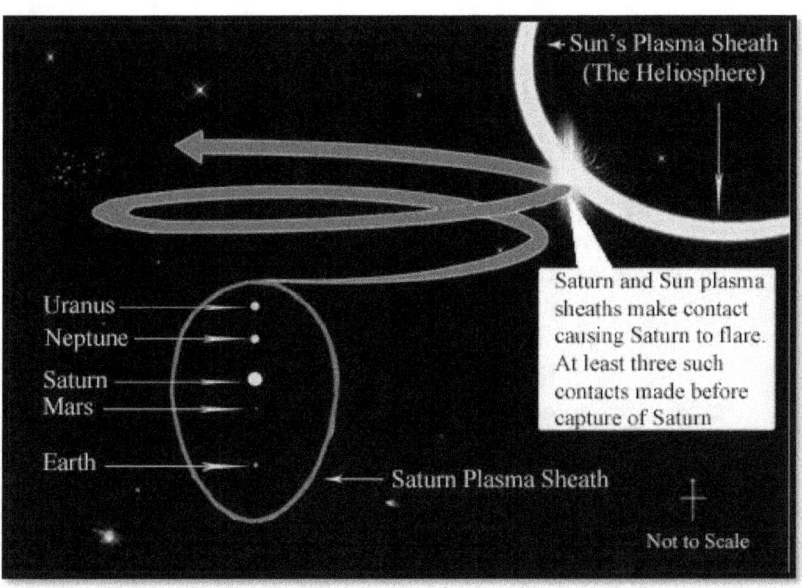

Our 9.5 conjecture says that the three final near approaches were critical and involved approaches that were close enough to allow for the possibility of humans living in one system ending up in the other: i.e., that people native to the Sun/Jupiter system had transferred to the Saturnian system. A more detailed sequence of events that outlines this scenario is that:

- The first of those final three encounters between the Sun's heliosphere and Saturn's plasma sheath caught any humans living on Ganymede unawares.
- A number of humans were transferred, possibly along with other species, to Earth by some means at that time.
- As the Saturnian system spiraled back out into deep space, any hope of those Cro-Magnon people returning to Ganymede was lost, leaving them stranded as castaways, strangers in a strange land.
- When the Saturnian system next approached our present sun, some thousands of years had elapsed with humans on Ganymede having attained an advanced level of civilization benefiting from thousands of years of technological development — they were now a fully-fledged space faring civilization.
- The Mars/Cydonia civilization was established by

267

Ganymedian space exploration and possibly even evacuation-themed expeditions during this second contact event between the plasma sheathes of the two planetary systems.

- The Mars/Cydonia civilization lasted 100 – 300 years as the Saturnian system's spiral towards the sun accelerated due to the phenomenon of electron acceleration taking place at inter-stellar scales, pulling Saturn, Mars and Earth towards a final and highly destructive third contact and capture event.
- A third contact between Saturn and the Sun's electrical fields was initiated leading to the capture of Saturn and its accompanying satellites, the flaring of Saturn into a fully-fledged sun and the destruction of the Mars/Cydonia civilization, while also bringing to a close the Pleistocene ice age on Earth.
- Life on Ganymede suffered an extinction level event as Jupiter was pushed out to its current frigid orbit as a consequence of the full capture of Saturn and its satellites.
- Refugees escaping Ganymede's extinction level event arrived on Earth as the forerunners to the biblical Adamic civilization at some time after Saturn's flaring and capture by the Sun.
- The third of those final three encounters resulted in the destructive, yet permanent capture of the Saturnian system by our present Sun and initiated the era on Earth known in world mythology as the Golden Age.

We reject the idea of the Mars/Cydonia civilization dating from much earlier than that second-to-last near approach of the Saturnian system to our present sun. We do not believe that any civilization centered at Cydonia enjoying a space faring capacity would have allowed fellow humans to remain castaways on Earth for thousands of years without attempting some sort of contact aimed at rescue, or at least bilateral relations.

We mentioned in the section titled **Caveats** that we were using the term "Cro-Magnon "in something of a nonstandard manner and we've just now seen what that means. There may have been more than one saltation of modern humans on this planet prior to the arrival of the well-known Biblical antediluvians and the study of haplogroups doesn't appear to shed much light on this question.

Some authors in fact believe that the people who created the *Venus of Willendorf* and similar art objects could not have been the same people who created the statuettes normally termed Cro-Magnon seen in a previous chapter here.   Nonetheless we are generally using the term"Cro-Magnon"as a catchall for any and all human groups that came to this planet prior to Adam and Eve and therefore would be associated with the near approach of the ancient Saturnian system to our present sun prior to the final and similar approach associated with the arrival of Adam and Eve. In other words, we are using the term "Cro-Magnon" where scholars 200 years ago would have used the term "Pre-Adamite."

As mentioned elsewhere in this work, humans living on Ganymede or Mars would have had no way of knowing whether anything in the system would remain habitable at all in the aftermath of Saturn's capture, although it seems likely that they would have viewed Earth, being the largest of the bodies in the system capable of supporting human life, as providing the best chance for survival in a habitable environment. However, for all intents and purposes, they appear to have relocated to places unknown, either to the near stars or to find an alternative permanent home in deep space. Either that or they too succumbed during the conflagration marking Saturn's entry into the Solar System.

Summary and Takeaways from this Chapter

Because of at least one recent population "bottleneck," there is very little genetic variation amongst modern humans. Historically, there were two original groups of modern humans according to saltation: Cro-Magnon people, and the familiar antediluvian people of Genesis. Again the two groups were essentially the same genetically, but their cultures and technologies were completely different. Likewise the difference had nothing to do with what you would call race; both groups are capable of producing any of the colors or features found amongst modern humans.

There is no way to believe that the people of Genesis were descended from Cro-Magnons. There is a list of things that Jewish literature and the Bible would have to know about were that the case and such knowledge simply isn't there.

Prior to its outright capture by our present sun, the Saturnian system, which Earth was originally part of, had made several near passes to the Sun/Jupiter/Mercury system. At least the final two such passes prior to capture involved near approaches of the two systems after which the Saturnian system would swing back out into space; in other words the Saturnian system was spiraling towards and then back away from the system of our present sun.

The difference in the cultures and technologies of Cro-Magnon and Bible antediluvian people is thus explained as a lapse of some thousands of years between those last two near approaches before capture. In other words, regardless of the exact manner in which humans were transferred from the Jupiter Ganymede system to this planet, logic dictates that those transfers would have occurred at the times of those near approaches.

Conclusions and Stray Thoughts

Anybody who has followed this discussion to this point will realize that the dialogue on a number of topics and subject areas will have undergone substantial change. Among the conclusions that this book suggests are, minimally (in the sense of low hanging fruit), the following:

- NASA, ESA, and the Western world in general have no other safe assumption than that they are now in a race to Mars and to Phobos. There is a real possibility that one of our planet's bad actors might get there first; there is no way to predict what such a bad actor might find there, or what they might do with anything that they were to find.
- There is also an absolute imperative that we get to Mars, Phobos, and Ganymede if for no other purpose than exploring our own origins.
- The findings of this book predict a high likelihood of finding archaeological sites on Ganymede. The images noted in the appendix section on Mars amount to archaeological evidence already having been found there.
- The findings of this book predict that the rock structure of Ganymede will be dominated by pumice.
- The findings of this book predict that the ice on Ganymede will be fresh water ice.
- There is a possibility that something that we might find on Mars or Phobos might allow us to attempt to regain contact with descendants of those who escaped either to the near stars or to deep space.
- It is past time that our space agencies level with the American and European people regarding evidence of past civilizations on Mars, our own moon, and other parts of our system. If, as Richard Hoagland believes, official thinking involves the Brookings Institute study from 1965 and a

belief that the American people "can't handle it," then it is our own agencies that need to get over it. The American people handled all of the news, good, bad, and indifferent from World War II and they can handle whatever news comes from space exploration.

- Some kind of an official funeral needs to be held for the theory of evolution and particularly the idea of humans being descended from hominids. As we have observed, modern humans and hominids are not even from the same world.

- Likewise, much of the magic physics of the 19th and 20th centuries (relativity, Big Bang, black holes, dark matter, dark energy, a gravity-only cosmos, etc.) will require being jettisoned before 21st century science proceeds much further.

At the time when modern humans first arose (anywhere close to here at least), it was within a double solar system, that is, a Saturnian system that included our own present planet, and the Sun/Jupiter system including Ganymede. There is a subtle and yet huge observation in this picture, i.e., that there is a rough similarity between the creatures that originated within the two systems. In other words, dinosaurs and hominids may look strange enough to us at first glance, but they were all bilaterally symmetric, had two eyes, noses, ears, scales or fur, four legs (or two arms and two legs). In other words, what we **DON'T** see is the kind of situation that you observe in Star Wars, of multitudinous kinds of fantastic and intelligent creatures with no constraints as to morphology other than artists' imaginations. For that matter, the creatures of both ancient solar systems are based on the same DNA/RNA information code.

Moreover, for all intents and purposes, the Sun/Jupiter system and the Saturnian system might as easily have been hundreds or thousands of light years apart. If for some perverse reason a person wanted to go on believing that the creatures of both systems have evolved in Darwinian fashions, he/she would have to explain how the same RNA/DNA information code and the same basic schemes for various kinds of creatures arose **TWICE** in two separate solar systems by random chance. There is also the related question of

Rupert Sheldrake's claim of "morphological fields" and the possibility that the various kinds of creatures that have inhabited our planet represent cosmic archetypes. What we have seen would suggest that the kinds of creatures which have lived on this planet both presently and in the distant past, are more or less the only kinds of creatures that we would ever encounter, no matter how far into deep space we were to travel.

If, on the other hand, a person is more inclined to believe that complex creatures were designed and created, then what we observe is roughly what we would expect. In other words, you would hardly expect an original creator to have designed and created modern man on and for a world for which he was stunningly ill adapted. What you **WOULD** expect is that the world for which modern man was created would be a world for which he was quite well adapted and, as we have seen, Jupiter's moon Ganymede was originally such a world.

Dolphins are mentioned along with humans in articles describing creatures with the proportionally smallest eyes. It strikes us as a fairly safe bet the Dolphins also were originally from Ganymede and that there were probably symbiotic relationships between humans and dolphins in those days.

Selected Bibliography

(Note: All periodicals, newspaper and online articles as quoted in main text)

(Books)

Ann Gauger, Douglas Axe, Casey Luskin, "Science and Human Origins." Discovery Institute Press, 2012, (Kindle Version)

Bakker, Robert T. "The Dinosaur Heresies"

Butterworth, E. A. S., "The Tree at the Navel of the Earth," (Berlin, 1970),

Cardona, D. "God Star," *Trafford Publishing*, Victoria B.C. Canada, 2006

Cardona, D. "Flare Star," *Trafford Publishing*, Victoria, B.C., Canada, 2007

Cardona, D. "Primordial Star," *Trafford Publishing*, Victoria, B.C., Canada, 2008

Cardona, D. "Metamorphic Star," *Trafford Publishing*, Victoria, B.C., Canada, 2011

Cohen, I.L., "Darwin Was Wrong - A Study in Probabilities," New Research Publications, 1984

Deloria, Vine. " Red Earth, White Lies: Native Americans and the Myth of Scientific Fact," Fulcrum Inc. (2000)

Desmond , Adrian J. "The Hot-Blooded Dinosaurs: A Revolution in Paleontology," New York, 1976

Fagan, B. "Cro-Magnon," Bloomsbury Press, (Kindle Version)

Hoagland, R.C. "The Monuments of Mars: A City on the Edge of Forever," Frog Books (2002)

Jaynes, J. "The Origin of Consciousness in the Breakdown of the Bicameral Mind," Houghton Mifflin, 2000

King James Version "Holy Bible"

Morgan, Elaine "The Aquatic Ape Hypothesis" Souvenir Press, (Kindle edition)

Nielson, Knut "Scaling, Why is Animal size So Important", Cambridge University Press, 1984

Remine, Walter "Biotic Message"

Ringwald, Frederick A. "SPS 1020 (Introduction to Space Sciences)" (2000)

Scott, Donald E. "The Electric Sky," Mikamar Publishing, 2006

Sheldrake, R. "The Presence of the Past: Morphic Resonance and the Habits of Nature," Icon Books, (2011 Kindle version)

Sheldrake, R. "The Science Delusion," Hodder & Stroughton, 2012 (Kindle Version)

Talbott, D. "Guidelines to the Saturn Myth." *KRONOS X:3* (Summer 1985)

Thaxton, Charles B. "The Mystery of Life's Origin: Reassessing Current Theories", Philosophical Library, 1984

Vendramini, Danny "Them and Us", Kardoorair Press (2009), (Kindle edition)

Wallace Thornhill and David Talbott, *"The Electric Universe,"* Mikamar Publishing, 2007

Wallace Thornhill and David Talbott, *"Thunderbolts of the Gods,"* Mikamar Publishing, 2005

(Selected PDFs, etc.)

Anto Tri Sugiarto, Masayuki Sato, "Pulsed plasma processing of organic compounds in aqueous solution," *Department of Biological and Chemical Engineering, Faculty of Engineering, Gunma Uni_ersity, 1-5-1 Tenjin-cho, Kiryu-shi,Gunma 376-8515, Japa,* Received 17 June 1999; accepted 12 July 2000

Bland, Michael Thomas, "The Tectonic, Thermal and Magnetic Evolution of Icy Satellites", *Phd dissertation submitted to the Faculty of the Department of Planetary Sciences, The University of Arizona.* (available online)

Bennett, Donahue, Schneider, and Voit; "The Cosmic Perspective," 5th Edition, Chapter 13.3, online PDF version

Adam Burrows, W.B. Hubbard, J.I. Lunine, James Liebert, "The Theory of Brown Dwarfs and Extrasolar Giant Planets," 2001, arXiv:astro-ph/0103383v1 , PDF

Geoffrey C. Collins, et al, "Ganymede science questions and future exploration," *Planetary Science Decadal Survey Community White Paper*, PDF

Kokh, Peter; Kaehny, Mark; Armstrong, Doug; Burnside, Ken. "Europa II Workshop Report". *Moon Miner's Manifesto* (November 1997)

M. G. Kivelson, K. K. Khurana and M. Volwerk, "The Permanent and Inductive Magnetic Moments of Ganymede," PDF, Icarus 157, 507–522 (2002)

R. T. Pappalardo, K. K. Khurana, and W. B. Moore, "The Grandeur of Ganymede: Suggested Goals for an Orbtier Mission" Brown University (Dept. Geological Sciences, Providence RI 02912-1846; 2UCLA (Los Angeles, CA 90095-1567), 4065.pdf

Solomonidou, A., Coustenis, A., Bampasidis, G., Kyriakopoulos, K., Moussas, X., Bratsolis, E., Hirtzig, M., "Water Oceans of Europa and Other Moons : Implications For Life in Other Solar Systems," *Journal of Cosmology, 2011, Vol 13. 4191-4211*

Appendix A, Telepathy and Pre-Flood Language

The question of antediluvian language does not fit the narrative flow of this book. Moreover, our view of the reality of this topic is so far from the beaten path that the general level of academic robustness which we have attempted for the book could not be achieved in this area. Nonetheless, we are quite comfortable with our own view on this question. It does relate to the question of how Cro-Magnon man appeared to be capable of passing information (new ideas and techniques, new materials) around over large distances before pony express systems or any other sophisticated system of communications existed. It also bears upon the question of whether the difference between humans and hominids had anything to do with language capabilities or whether Neanderthals were capable of speech as we know it or whether they needed speech as we know it. What follows therefore is our own vision of what pre-flood language amounted to, caveat lector…

Isolate Languages

We mention in another part of this work that evolution works for small things but not for large things, that is, that microevolution is sufficiently real, but that macroevolution is not. Numerous experts are on record to the effect that microevolutionary changes cannot agglomerate into anything that you would call macroevolution[6]. The situation involving the evolution of human languages is entirely similar. Evolutionary change can account for the difference between our English and Chaucer's, or for the difference between

---

[6] E.g. Prof. R. Goldschmidt, - PhD, DSc Prof. Zoology, University of Calif. In Material Basis of Evolution Yale Univ. Press, "It is good to keep in mind ... that nobody has ever succeeded in producing even one new species by the accumulation of micromutations. Darwin's theory of natural selection has never had any proof, yet it has been universally accepted."

Ukrainian and Russian, but there are simply too many things that cannot be accounted for.

Total isolate languages like Spanish Basque or the language of the Japanese Ainu or the multitudinous languages of Australian aborigines are examples of language-related issues that language evolution cannot account for. Another problem would be the Baltic languages. . .

Consider Lithuania, which lies directly between the Germanic and Slavic worlds. The people have blonde hair and blue eyes, and one might assume the language would have to be half way between German and Russian; in real life, Lithuanian appears to have a couple of dozen elements that you might call Indo-European roots and the rest of the language appears to come straight from some other solar system. English is much closer to Russian than Lithuanian is.

But the biggest problem for anybody wishing to believe in any sort of macro evolution for languages is the Indo-European/Semitic divide. There is no meaningful racial difference between the two groups and there is no reason to believe that they might have split up more than a few thousand years ago. You would expect the two groups of languages to be strongly related and, yet, other than for a few borrowed words, they do not appear to be related at all.

To get some idea of how many things remain recognizable between Indo-European languages, let us imagine for a moment that a person who spoke only English were to decide to make Russian his first foreign language to study. In other words, rather than studying a language which split from yours 1000 or 1500 years ago as would be the case of an English speaker studying German, let's see what happens when you go to study a language which split from yours two or three thousand years ago. This person would certainly experience enough pain with the system of declensions and verb aspects, but he/she would find a startling number of things that were familiar:

- Numbers: nearly all the same or at least recognizable other than for 'devyat' (nine).

- Family members: nearly the same. 'Mats/mati' (mother) declines as 'materi'; 'ahtyetz' (father) amounts to the same 'aht' in 'father' or 'pater', and the 'yetz' part of the word is a generic suffix.
- Personal pronouns: nearly the same or at least recognizable in all cases.
- Common things: fire ('ah-gon', like ignire/ignition/Agni); water ('vod-ah'); wine ('veen-oh'); wind ('vyeter', like vent, ventilator etc.)
- P/F words, which start with a p in one IE language and with an f in the next: flame/plamiya, fall/pahl (упасть, попасть, пропасть, etc.) , flow/plavats(swim) Familiar examples from more familiar languages include foot/pied/pedal, fish/pesh etc. These words arise because Indo-Europeans originally pronounced P's and F's together. German retains words like that, e.g. pfennig/penny or pferde/horse.
- D/G words: give/davats gimme/дай-мне/dai-me (Slavic kid language)
- Common household things: knife/nozh, spoon/ladle/Loeffel/lozhka, kettle/kot-yohl etc.
- Old/very-old IE roots: it (step)/идти/iterate/itinerary
- Feudal relationships: dolg/dolzhen (debt, obligation) = do + L-G (as in liege lord), delegate, relegate, obligate, allegience etc.

Thus, despite the family groups for English and Russian having split up two or three thousand years ago, there is a great deal that remains similar between the two and, in fact, appears likely never to change beyond recognition no matter how many centuries might pass. By all rights you'd expect some of these similarities to also exist between Indo-European and Semitic languages, but they simply don't.

What the actual evidence suggests is that human communication was originally of some completely different nature until some very recent point at which whatever that consisted of suddenly stopped working on a single day, and the kinds of spoken languages that we use today were thereafter devised very rapidly out of dire necessity. **GIVEN** that hypothesis, all which would be needed to explain the

Indo-European/Semitic divide would be that the two groups were separated by the Caucasus mountains for that critical period of one or two centuries during which spoken languages were being devised. Likewise Lithuanians, isolated in their forests, devised their own language while picking up words here and there from people passing through.

Back to the question of Cro-Magnon people...

More than one author has noted that ideas appeared to spread extraordinarily quickly amongst Cro-Magnon people. Remember that these people were supposedly on foot with nothing resembling reasonable communication technology whatsoever. This was before anybody even thought of domesticating horses, much less creating anything resembling Kublai Khan's pony express (yam) system.

Laura Knight ("Secret History of the World") notes that:

> "But it seems that right from the beginning, Cro-Magnon man was traveling and sharing and exchanging not only goods, but technology.
>
> If there was a better form of stone somewhere else, the word seemed to get around, and everybody had some of it. Distinctive flints from southern Poland are found at Dolni Vestonice, a hundred miles to the south. Slovakian radiolarite of red, yellow and olive is found a hundred miles to the east. Later in the Upper Paleolithic period, the famous "chocolate flint" of southern Poland is found over a radius of two hundred and fifty miles."[7]

Cro-Magnon people were in fact Stone Technology people and there simply isn't any way to make a radio out of stone. Nonetheless, the evidence suggests that they may in fact have had access to a global

---

[7] Laura knight, "Secret History of the World (http://tinyurl.com/c7thl9w), page 237

system of communications that did not require radios. An investigation of the question of telepathic communication in prehistoric times leads to the works of two very excellent scholars: Julian Jaynes, and Rupert Sheldrake.

The Evidence for Telepathy in Today's World

The original modern use of statistical science appears to have involved determining the effectiveness of crop treatments and similar problems in which direct measurement was not practical or where it was not possible to take all relevant variables into account.

Rupert Sheldrake is a former director of studies for cell biology at Cambridge University who, in later life, has taken up the study of things normally termed "paranormal," using good experimental design and statistical methodology. Thus his use of statistical methods to investigate the phenomenon of people guessing correctly most of the time whether or not they are being stared at by others is entirely in keeping with the manner in which the science of statistics is meant to be used. In the case of dogs who appear to know the first moment their owner heads for home (some dogs are observed going to sit at the door when their owners begin their journeys home), Sheldrake has devised a simple and elegant test. He has had dog owners set out on window-shopping tours along with one of his (Sheldrake's) assistants with only the assistant knowing the predetermined time to start home. Sure enough, in such cases when the owner is told that the time to return home has arrived, the dog has dutifully gone to the door as noted by another assistant who would remain at the owner's home.

By these and other methods, Sheldrake has demonstrated to a statistical certainty that a number of things normally designated as paranormal, are real. Naturally enough this has earned him the animosity of the CSICOP[8] crowd and of other professional skeptics.

---

[8] Committee for the Scientific Investigation of Claims of the Paranormal (CSICOP), now called Committee for Skeptical Inquiry (CSI)

Sheldrake has also noted that major league computational power, until recently, was an exclusive prerogative of government agencies, large corporations, and large universities, but that the common man now has access to such computing power and, with it, significant scientific investigation on the cheap. His "Seven Experiments that Could Change the World" is a book that everybody should have.[9]

And there is no shortage of his materials on YouTube e.g.

http://www.youtube.com/watch?v=JnA8GUtXpXY

Based on a number of Sheldrake's results, there is thus substantial reason to believe that there is at least some kind of a minimal telepathic capability amongst humans and other higher animals even today. But what about the past? Is there reason to believe that what Sheldrake describes is a **REMNANT** of a much more significant telepathic capability from past ages?

Telepathy in the Ancient World

That question leads us to the works of Julian Jaynes, a psychology professor and amateur philologist at Princeton University whose main interest in historical questions involved Homer's *Iliad* and other literature from the same time period. Reading through such materials, Jaynes began to notice the curious absence of decision making in the *Iliad* and in other works from the same age. At every point at which you or I would have to stop to consider how to proceed, the people in these ancient narratives are being told what to do by inner voices, which are described as gods and goddesses.

It began to dawn on Jaynes that what we would call schizophrenia today, hearing voices, was likely the normal state of affairs in ancient times. At that point, Jaynes appears to have gone to the experts in neurophysiology at his school with the obvious question: "What if anything could there possibly be in the human brain that would cause somebody to hear voices for no apparent reason?" The basic answer was that there actually is a right brain analog to the speech center (Wernicke area) on the left side of the human brain. For all intents and purposes this right-brain analog of the speech center appears to be a neurophysiological equivalent of the human appendix, i.e. it serves no apparent purpose. However, when this right side analog area is stimulated with electrical probes, as is sometimes done in experiments with epileptics, more often than not the patients claim to be hearing voices as real as if someone were speaking to them.

All of this motivated Jaynes to write a book with the impossible title: "*The Origin of Consciousness in the Breakdown of the Bicameral Mind*", which became an academic sensation in the mid to late 1970s.[10]

Jaynes was an evolutionist; he assumed he had discovered that, at least during the time between Exodus and Alexander, which was the time frame he was interested in, human societies had simply evolved into a state in which they were being controlled by what he termed "auditory hallucinations." Thus his diagram (reproduced below) that shows the speech center, the right side analog to the speech center, and the bridge crossover between the two, and denoting the right side analog to the speech center as a "hallucinatory" area.

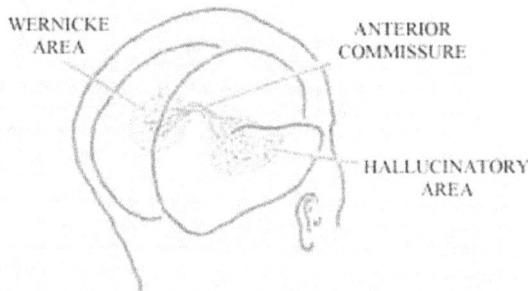

(Image from Jaynes "Origin of Consciousness," page 104)

The introduction to this work mentioned that ancient literature involves descriptions of things outside of our experience, including:

---

[10] http://www.amazon.com/Origin-Consciousness-Breakdown-Bicameral-Mind/dp/0618057072

Descriptions of religious practices intended to communicate directly with the spirit realm. Such practices included prophecy, oracles, "familiar spirits" (the tale of Saul, Samuel, and the witch of Endor, etc.), idolatry and the rituals associated with the worship of idols, and electrostatic devices such as the Ark of the Covenant. All such practices involved trance states similar to hypnosis, all involved static electricity, and they all stopped working prior to the time of Alexander.

Jaynes' work *"Origin of Consciousness "* is largely an attempt to explain those kinds of phenomena. But he neglects the history of the world prior to the books of Moses, i.e. the question of Genesis, and the question of the origin of language. The words "prophet" and "prophecy," for instance, which permeate the books of the Old Testament after Genesis, do not occur in Genesis other than the one vague reference to Abraham as "God's prophet," which occurs after the flood and the incident involving the tower at Babel.

Most assume that the tale of the tower of Babel amounts to a claim that men originally spoke one language such as we speak now and that God then caused them to speak multitudinous languages such as we speak now. Is that the case, or could it be that some larger change is indicated? Could it be that the phenomena that Jaynes described formed the basis for the original human communication capacity? Could it then be that, for some reason that stopped working, and that languages such as we now speak were devised thereafter?

The *King James Bible* notes that:

> GEN 11:1 And the whole earth was of one language, and of one speech.

At least one excellent Hebrew language scholar, Lisa Beth Liel, claims[11] that the passage has been mistranslated and that:

---

[11] Personal correspondence.

"Safah achat = "one language", "devarim achadim = "few words."

In other words, the phrase should read: **"One Language, Few Words,"** or possibly, "One language, few **SPOKEN** words." This appears to mean that before the incident associated with the tower of Babel, there was a minimalistic set of spoken words used for ritualistic purposes, but not as a general system of communication. Egyptians referred to these words as "words of power." Dr. E.A. Wallace Budge describing the Egyptian God Thoth, who was called *was called "lord of divine words" and "mighty in speech"* :

> *". . . from one aspect he is speech itself. . . Thoth could teach a man not only words of power, but also the manner in which to utter them. . . . The words, however . . . must be learned from Thoth."[12]*

Julian Jaynes assumed that human societies had simply evolved into a state in which they were being controlled/governed by systems of "auditory hallucinations" that were experienced by groups as well as by individuals. He assumed that this sort of "bicameral mind" passed away and that what we call consciousness developed due to societal changes in a manner consistent with ideas about evolution.

Nonetheless, we see overwhelming problems with the idea of man evolving into what amounts to a dysfunctional state — and the world of the Old Testament was intensely dysfunctional. Fighting wars and sacrificing children at the behest of stone or wooden idols is not a formula for success in life and if any one group of people were to start living that way, they would be at a gigantic evolutionary disadvantage versus every other group of people who did not. We see no way to believe that such a way of living could become dominant had it not been a primordial condition. Likewise, assuming that such a state had somehow become the common

---

[12] The Gods of the Egyptians (London, 1904), vol. I, p. 401; cf. P. Boylan, Thoth the Hermes of Egypt (Oxford, 1922) and B. von Turayeff, "Zwei Hymnen an Thoth," Zeitschrift fuer Aegyptische Sprache 33 [1895], pp. 120-125.

condition of mankind, we see no way to believe that anybody would find a gradualistic way out of such a state of affairs.

What we believe to be the case is that:

- Communication amongst humans, and most likely amongst higher animals as well, was originally telepathic and that this capability depended in some manner upon the electrostatic environment of the ancient solar system.
- At the time of the incident associated with the Tower of Babel, whatever that previous telepathic communication system depended upon broke down, and has never been restored since.
- Humans, with their voluntary control of breathing, were able to devise spoken languages at that time, but other creatures, which previously had access to complex communication capabilities, have lacked them since that time.
- The human brain was somehow rewired in that process.
- Because the ancient telepathic communication system involved a part of the human brain that still exists, the capability would have been primordial to the human race, that is, Cro-Magnon people would have possessed it.
- The religious phenomena that Julian Jaynes describes (oracles, prophecy, idols/idolatry, familiar spirits, etc.) represented remnants of the former system, which survived for a certain time and then stopped working altogether prior to the time of Alexander.

We believe that Julian Jaynes was correct in describing the voices that people heard coming from idols as hallucinations; there is no way to believe that anything good ever could have come of harking to the voices of stone or wooden idols. The cases of oracles, prophets, and "witches" (those with familiar spirits) are more complicated. Entire nations appeared to have run for centuries on information coming back from the trance states of oracles and prophets. Even Jaynes noted that it was hard to picture any society operating in such a manner if **ALL** such information had been garbage. But at some point all of those practices ceased working. The information turned to mush, and it became dangerous to listen to it. Thus from Exodus:

> EXODUS 22:18. Thou shalt not suffer a witch to live.

And later, Zechariah:

> ZEC 13:3 And it shall come to pass, that when any shall yet prophesy, then his father and his mother that begat him shall say unto him, Thou shalt not live; for thou speakest lies in the name of the LORD: and his father and his mother that begat him shall thrust him through when he prophesieth.

In other words, thou shalt not suffer a **PROPHET** to live either!

Thus, presumably somewhere around the time of the Trojan War, man's ancient communications capability came to a final end and the last vestiges of it died out. In the absence of other information, we would assume that going backwards in time from the biblical flood to the earliest point at which modern humans came into existence; mankind always had this proposed ancient capability.

One final note here... **IF** Lloyd Pye is right (and that is a gigantic 'IF'), and there actually are small numbers of remnant hominids walking around in the 65% of the planet's land surface that is seen only from the air, then the following problem would arise. You would have to note that some of the reports originating in the 1800s indicated fairly sophisticated hominid ("archaic") burial sites while today's "Sasquatch" is generally described as more of a wild animal. That would in fact be what you might expect, assuming that such creatures had sophisticated communications capabilities in prehistoric times, but have not had them for the past 3000 or 4000 years.

Appendix B, Mars

This appendix amounts to another piece of background information which does not fit into the narrative flow of this book. It is included here to provide a flavor of the kind of information which has been coming back from NASA and ESA Mars probes since 1999, and of the controversies surrounding some of the images in question.

Faces

The first hint that places in our system other than our own planet might have been inhabited in prehistoric times came in 1976 with the Viking probe and its photograph of the familiar 'Face on Mars' seen in Mars' Cydonia region:

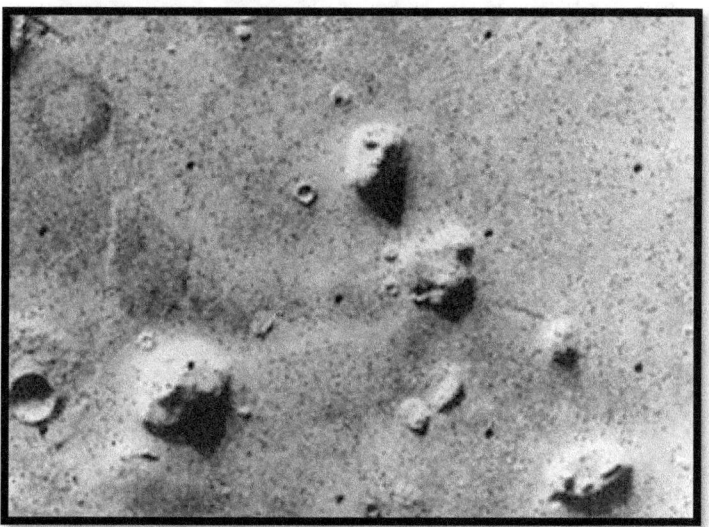

(Face on Mars, Original 1976 Viking Probe low-resolution image)[1]

The 'Face on Mars' appears to resemble a gigantic megalithic structure in the form of a human face, approximately 1.2 miles by 1.6 miles, meant to be viewed from on high or even seen from off planet. NASA of course claimed (and still claims) that the structure is a natural formation. What most people are still unaware of is that many more detailed images started coming back around 1998,

---

[1] http://www.stoptherobbery.com/Space.html

including astonishing pictures of pyramids in the Cydonia region and this new image of the face megalith:

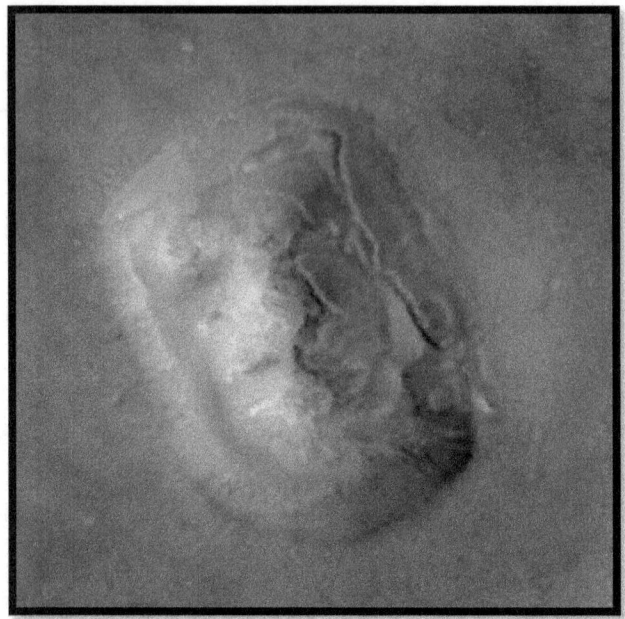

(2001 MOC high-resolution image)[2]

That image should have ended any debate on the topic; Mother Nature does not create straight lines and Bezier curves on a three-mile scale. Nonetheless the official position of the various space agencies remains that all such images are natural formations. Since then, other Martian face images, made to be seen from off planet, have been found. Dr. Thomas Van Flandern, a former director of the Naval Observatory, provided proof of the artificiality of the Cydonia face megalith.[3]

His website, www.metaresearch.org, includes slides from a press conference he held in 2001 that dealt specifically with the Mars image findings:

http://www.metaresearch.org/solar%20system/cydonia/asom/artifact _html/default.htm

---

[2] http://mars.jpl.nasa.gov/mgs/msss/camera/images/moc_5_24_01/face/
[3]
http://metaresearch.org/solar%20system/cydonia/proof_files/p roof.asp

The slides presented by Dr. Van Flandern include images of other human faces seen on the Martian surface, which are reproduced below.[4]

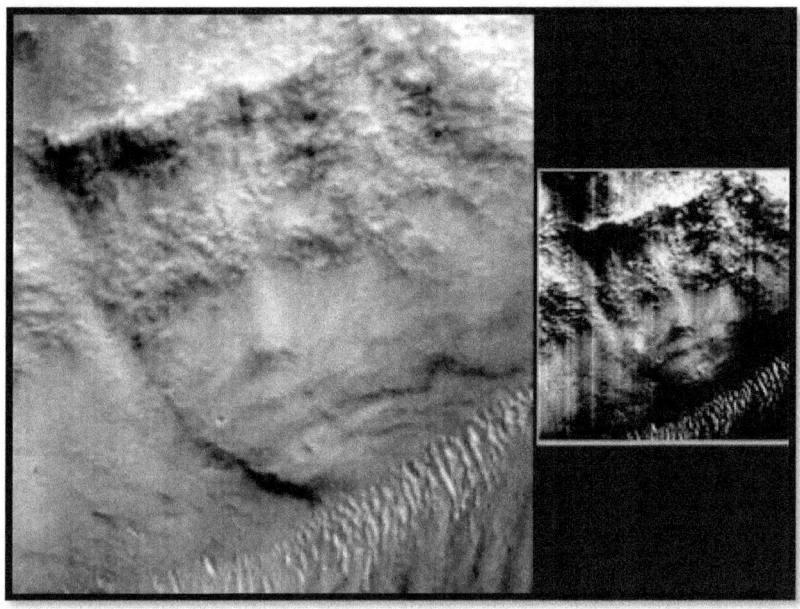

(Van Flandern 2001 Press Conference, Slide 44)

4

http://www.metaresearch.org/solar%20system/cydonia/asom/artifact_html/
default.htm

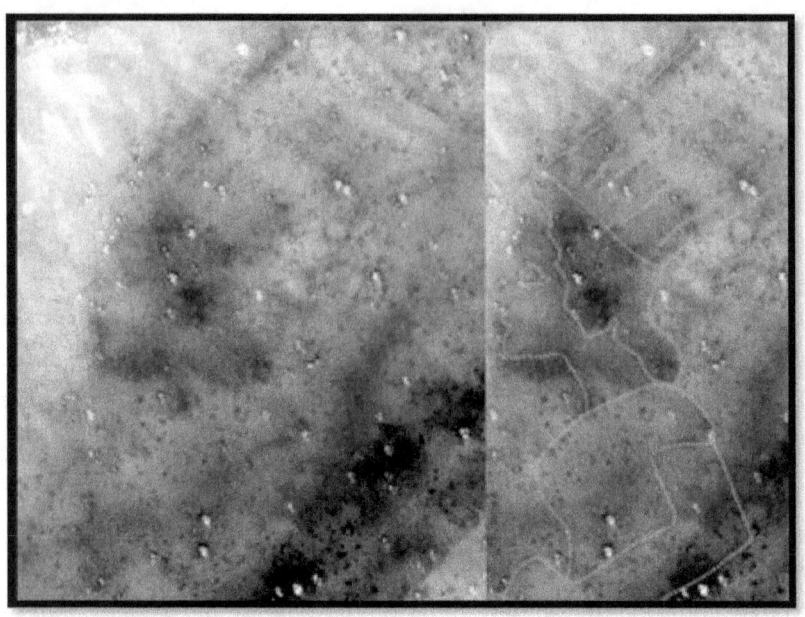

(Van Flandern 2001Press Conference, Slide 42)

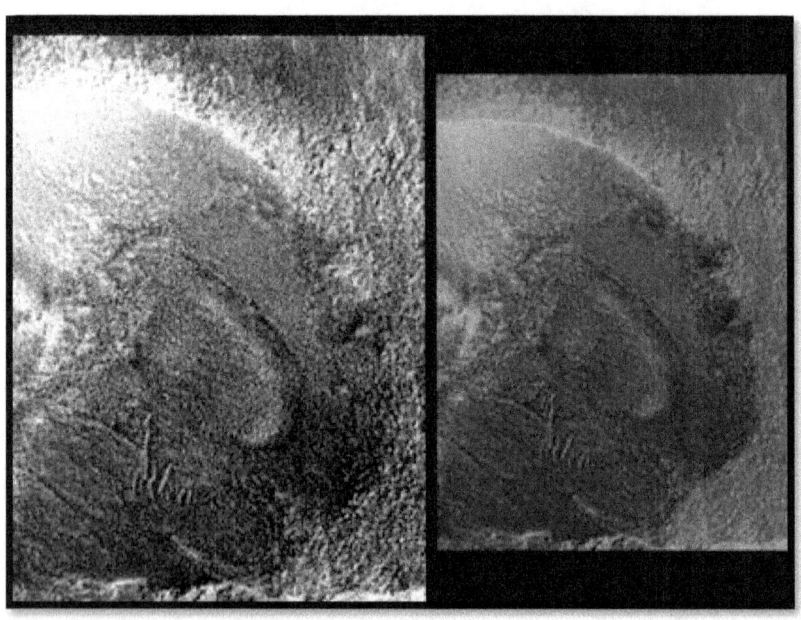

(Van Flandern 2001Press Conference, Slide 43)

These faces are clearly those of modern humans and not hominids. There are also images depicting animals, that is, images of structures representing animals that can also be seen from off planet.

Van Flandern's demonstration of the artificiality of the face megalith at Cydonia amounted to demonstrating that it passes the basic fractal test for artificiality. In simple terms, there are natural formations (clouds, mountains etc.,) which appear to resemble some artificial or living thing. Invariably, however, the resemblance diminishes as the viewer approaches more closely. The face megalith, on the other hand, looks MORE artificial with increasingly high resolution views.

Pyramids

There are a number of artificial structures in the same area as the face megalith including pyramids. One of those is truly gigantic and five-sided:

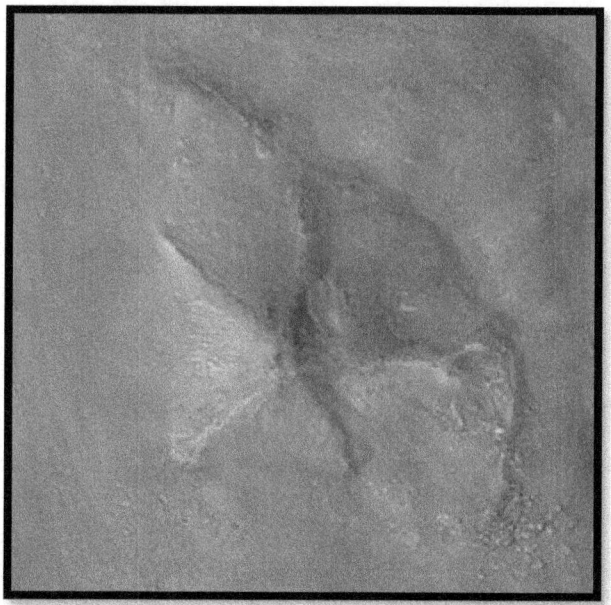

Cydonia's D&M Pyramid, long sides approximately 1.5 miles, Malin images at http://www.msss.com/mars_images/moc/2003/09/15/)

We've mentioned that nature doesn't do straight lines or Bezier curves on three-mile scales; nature doesn't do five sided pyramids with buttressed corners, a mile and a half on an edge either.

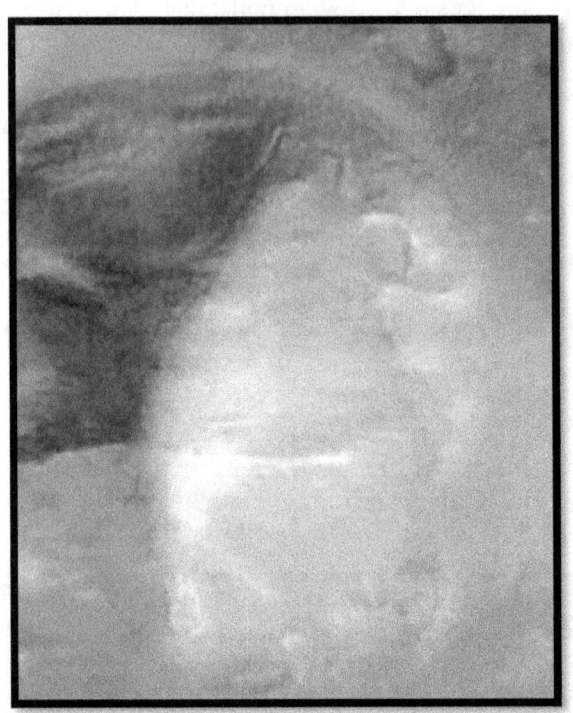

Topographical view of Cydonia's 'Main City Pyramid.[5]

The Mars Anomaly Research website has numerous Martian pyramid images.[6]

Cities

More recently, images showing what appear to be the remains of vast cities have appeared in NASA and European Space Agency(ESA) images. That sort of evidence is in no short supply on YouTube and on the website www.marsanomalyresearch.com. In fact ESA publishes these images because their mandate requires them to do so. But first they attempt to eliminate the anomalous-looking material by fiddling with the brightness and contrast of the images to make them look like sandy desert areas. Sharp-eyed private researchers and bloggers have found several of these areas

[5] MOC images:
http://www.msss.com/moc_gallery/ab1_m04/images/SP125803.html
[6] http://www.marsanomalies.com/pyramids

and removed the blurring. The authors have performed this exercise
two or three times to convince themselves of the content of these
images. The simplest example is the case of the Hale Crater
(pictured below):

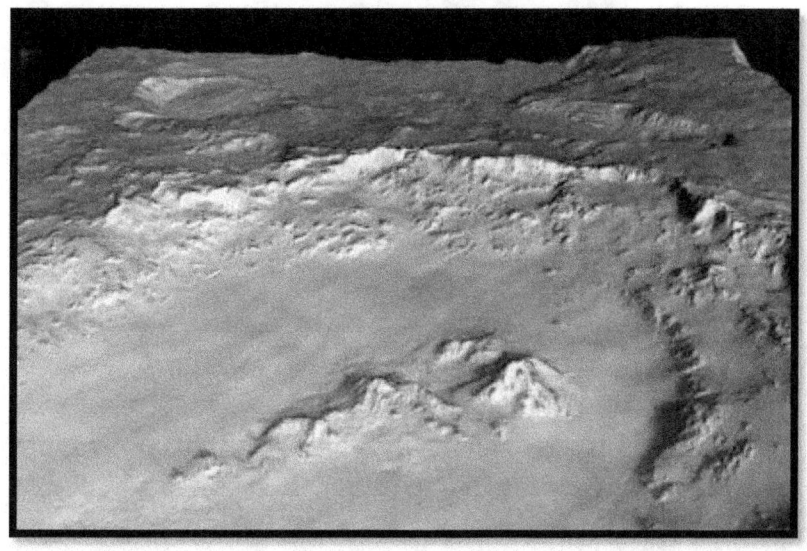

Picture credit:
http://spaceinimages.esa.int/Images/2004/11/Crater_Hale_in_persp
ective_looking_west

Just a sandy wasteland, right? You can download the high
resolution jpeg image at the following link for those wanting to
investigate our claims further:

http://esamultimedia.esa.int/images/marsexpress/137-021104-0533-
6-3d2-01-HaleCrater_H.jpg

Feed that image into any decent image handling software, such as
the Free Software Foundation's Gimp Package, adjust brightness
down and contrast up a bit and, voila, the image doesn't look like a
sandy desert any more:

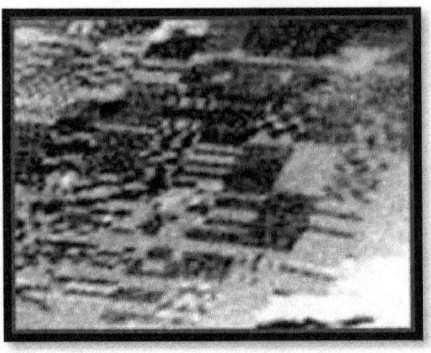

Picture credit:http://www.marsanomalyresearch.com/evidence-reports/2005/084/hale-civ-evidence.htm

Another ESA image area that can be retrieved from showing a desert by adjusting color hues, is reproduced below:

An explanation of the derivation of this image is available on Youtube.[7]

Again, the original ESA image looks like a desert area.[8]

---

[7]

http://www.youtube.com/watch?v=2DoZHWi1_oA&feature=player_embedded
[8]

http://www.esa.int/Our_Activities/Space_Science/Mars_Express/The_Medusa_Fossae_formation_on_Mars#subhead3

Phobos

Recently also, color images of Mars' little moon Phobos have been published, and a simple Google search on "Phobos HIRISE" will turn up these 2008 HIRISE project images from the University of Arizona:

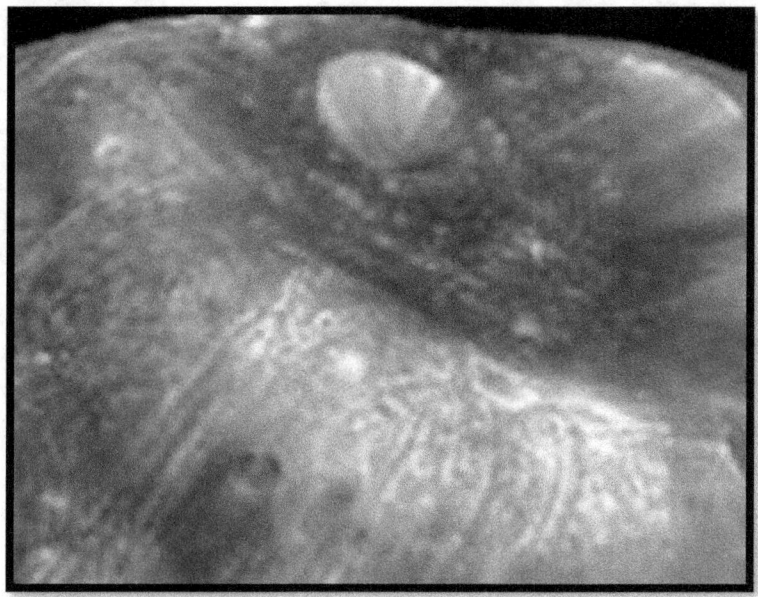

Real moons of course are supposed to be made out of dirt and rocks; they are not supposed to be made of material displaying metallic strakes or be able to reflect light all over the place like that. <u>That thing is artificial</u>, an ancient space station of some sort.   Readers eager to learn more about this topic are directed to Richard Hoagland's website.[9]

Other Artificial Structures

Structures that are clearly artificial have been found in other parts of our system as well but not in the profusion in which they are found

---

[9] http://www.enterprisemission.com/Phobos.html

on Mars, so we will not burden the reader with a cataloguing of these anomalies save to mention Saturn's little moon Iapetus:

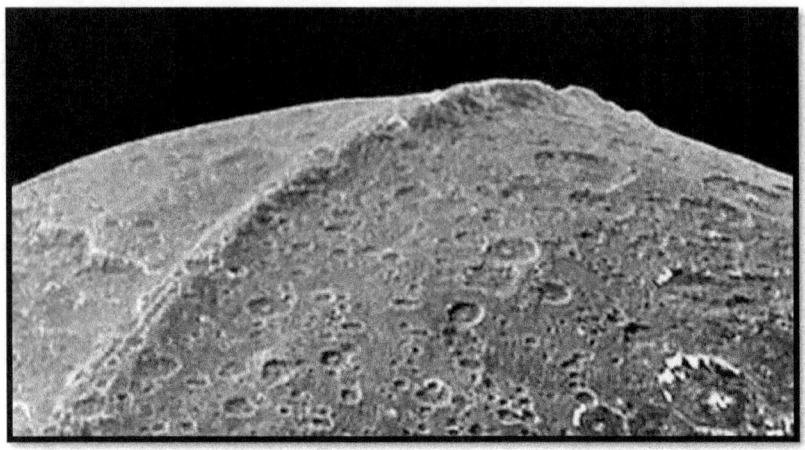

A shot of Iapetus' surface showing the curiously straight and high wall feature that circumnavigates this satellite of Saturn. Image credit: NASA

We've mentioned the fact that Mother Nature does not create straight lines or Bezier curves on three-mile scales. She doesn't create 9-km high walls on great circle arcs around moons and satellites either. So, despite all of the denial coming out of the space agencies, we view the proposition of there having been an advanced technological civilization in our system during prehistoric times as a virtual certainty.

Until humans get to Mars and spend the several decades it will take searching for artifacts and whatever records may have survived, most of what we will have in the way of a narrative is conjecture. In the case of Phobos, however, some of the conjectures that suggest themselves seem reasonable enough. A reasonable space station for instance shouldn't need to be 15 miles across; we assume that Phobos was constructed to serve some extraordinary purpose. A reasonable conjecture is that somehow or other, Phobos was associated with a great escape either to the near stars or to the cosmos at large, sometime after the inhabitants of Mars determined that their planet would not remain habitable after whatever was threatening them took place.

Again, the faces seen in those images carved into the Martian surface are those of modern humans and not of hominids or of any sort of space alien. That certainly does not support claims that modern humans evolved from hominids on this planet.

Reasons for Denial

As of the last I had ever heard [Ted speaking here] from anybody with actual NASA contacts, both NASA and the JPL were about evenly divided on the subject of Cydonia/Mars; about half of the people were willing to believe their own eyes and the other health didn't want to hear about it. The problem is that, given the standard theories of the history of our solar system, there is no way to believe that Mars ever could have been habitable; it would always have been too far away from the sun, too cold, and it would never have been able to hold a breathable atmosphere (at least via gravity). That is to say, in order to believe the evidence and what their eyes are plainly telling them, NASA and ESA scientists would have to completely scotch everything they thought they knew about the history of our solar system, go back to the drawing board, and start over.